A HISTORY OF ROSBORO LUMBER COMPANY

Design by Richard Warren
Printing by: Mira Digital Publishing
Printed in the United States of America
ISBN: 978-0-9800094-0-8

Library of Congress Cataloguing-in-Publication Data

Wells, Gail E., 1952–
A history of Rosboro Lumber Company: How a hardworking Oregon lumber company survived and thrived into the twenty-first century
ISBN 978-0-9800094-0-8
1. Lumber trade–Oregon–History. 2.Timber–Oregon–History. 3. Wood products industry–Oregon–History.
2007908537
CIP
This book is printed on acid-free paper.

A HISTORY OF ROSBORO LUMBER COMPANY

How a Hardworking Oregon Lumber Company
Survived and Thrived Into the Twenty-first Century

GAIL WELLS

Foreword by Paul Cole

DUNAWI CREEK PRESS

Acknowledgments

This history was commissioned by Rosboro Lumber Company, with the understanding, right from the beginning, that the first mission was accuracy. The world is full of company histories that are not much more than advertising and flattery. In contrast, this author was given a free hand to range through company archives, private letters, and the memories of company long-timers and tell the story she found. Companies are built and run by strong-minded human beings, and it would be surprising—indeed, unbelievable—if the path were portrayed as always smooth and the travelers always mild and agreeable. For not insisting on a whitewashed version of the truth, I am deeply grateful to the company's principals, chiefly Paul Cole, David Weza, and Jim Walsh, as well as to all who shared their memories with me. Thanks in particular to Wes Anderson, Irene Batey, Fred Bauer, Wes Bergen, Dave Blake, Nonnie Cole, Gordon Culbertson, Dallas Davis, Roy Edwards, Jim Enright, Ray Everts, Claude Gerlach, Roger Goodwin, Jon Howard, the late Clyde Hughes, Richard Johnson, Ruth Cole Kainen, Hub Lindsey, Ira Lowe, Knox Marshall, Louise McQuin, Scott Nelson, Margaret Rieder, Gene Scherffius, Avenelle Smith, Larry Taylor, Jim Walsh, Jim Watts, David Weza, and Larry Works. Thanks also to Nola Dennis, Sondra Jameson, and Jan Kessler for helping me make useful connections. My gratitude to Gordon Culbertson for furnishing many of the photos of company logging operations.

I am grateful to George Brown, Terry Brown, Paul Ehinger, Jeff Ernst, Rick Fletcher, Darrel Kenops, Grady Lewis, Robbie Robinson, Michael Thoele, and Zane Smith for offering their perspectives. I particularly thank Mike Thoele and Zane Smith for their thorough and perceptive reviews of the manuscript.

CONTENTS

STAND
ROSBORO
UTIL
ROSBORO
ECON
ROSBORO

FOREWORD

For some time it has been on my conscience that we simply had to compile and write down the history of this company. I wish now that I had begun about 20 years ago. Had this work started earlier, we would have had many more long-term veterans to draw upon for their memories and recollections of the company, its beginning, the early days, the war years, changes in our plants and operations. I suppose I really determined to begin this history after talking with my friend Ed Raggazino at a yearly cocktail party. He would get me to telling stories of the wood business, the people, and other factors. In all of these meetings, he would urge that we get a good writer to help us begin this work.

This year of 2007 marks the sixty-sixth year we have been in business. In some ways our story and that of our predecessor company, Caddo River Lumber Company in Arkansas, are intertwined. In 1906 Whit Rosborough founded Caddo River, and in 1939 he founded Rosboro Lumber Company. While not all stockholders and employees were the same in both companies, some were. To name a few: Whit Rosborough himself; my father, Beuford Cole; Vernon Williams; and the crew that came from the Caddo mills to build and work for the new company.

I hope by this book to explain how we got to where we are today and to show what we hope to be in the future. The past often indicates what future directions will be taken. Rosboro has always been a manufacturer of high-quality products, of which we are proud. We have always had a major timber base for our sawmill and other operations, and these plants have been important to our employees, our customers, and the business community.

Continuity has always been important to us, and now it is even more so. We pride ourselves on planning our future. As a major competitor once put it, "Rosboro has always been a well-planned company. You seem to know where you are going all the time."

Many sawmills, plywood plants, and glulam beam plants were here before us, and some new plants came into being after I began to work full-time for Rosboro. We are still here. We are proud not just of our survival but of our success. An operations manager once told me, "To employees who question what we do, I say, 'Look at our record. Rosboro must have been doing something right all these years.'"

When I was made general manager, I determined we should become an integrated manufacturing company. I also believed the company should be much larger than the one I began with. This led us into veneer and plywood in 1959, then glulam beams in 1964, and later it led us to build one of the area's first small-log sawmills. Many plant changes were made in buildings, equipment, and material flow. These steps and changes were all designed to yield us more profit, increase our efficiency, and give us a diversity of products. In the normal cycle of business, some of our products are up while others are down, and this has helped us spread the risk of our enterprise.

I have been fortunate in having a group of owners who let us accomplish what was necessary to make this a truly great company. All the changes I felt necessary to meet our long-time goals and bring our plants completely up to date required the approval and understanding of our board of directors and owner group. They have supported these efforts wholeheartedly.

Some people reading this history will be new to the company. Others will have been associated with it for many years—perhaps most of their working lives. Some of the history in this book will be familiar to them, but much will be new. I hope all will find it interesting.

We have been a successful company, and we look forward to a successful future.

Paul Cole
President, Rosboro Lumber Company

1

A LUMBERMAN COMES TO SPRINGFIELD

"If I am willing to put my life's earnings of seventy years of age in this business, no man should hesitate to do likewise."[1]

—Thomas Whitaker Rosborough

Before the stores and houses and sawmills were here, before the leagues of sportfishermen in their drift boats, before the paved streets, before the bridges, the streetcars, the railroads; before the filbert orchards and the hop yards; before the Missourians and Kansans trudging west over the Applegate Trail, even before the Kalapuya people, there were the trees. Douglas-firs and hemlocks 300 feet tall and 27 feet around the trunk, rearing up from thickets of vine maple and salmonberry, standing for four hundred years. And then, toppled by wind or fire, sliding into the rivers, decaying into duff, and making way for younger giants to succeed them.

The trees lined the canyons of the McKenzie and Willamette Rivers. Flowing cold and clear off the Cascade Mountains, the rivers embraced a fertile and pleasant land. The parcel of grassy bottomland at the confluence of the rivers seemed to Elias and Mary Briggs a promising place for a homestead. Traveling from Kentucky over the Applegate Trail, one of several branches of the Oregon Trail, Elias and Mary had arrived in the Willamette Valley in 1849. They took out a 640-acre donation land claim on the east bank of the Willamette River, a spot with a bubbling spring.[2] They named their new home, prosaically enough, Springfield.

The Briggses were followed by other pioneers from Kentucky, Missouri, Iowa, Tennessee—farmers, storekeepers, entrepreneurs. They came for the free land and a chance to get ahead in this moist, fresh country. Most prized of all was the grassy prairie land of the Willamette Valley, which, although it looked like a pristine paradise to them, had been kept mostly free of trees by the Indians' centuries-old practice of burning off the grasses in the fall.[3]

But the trees were also an attraction. There was alder and cottonwood along the river banks, and oak on the valley knolls, and Douglas fir and hemlock on the Coast Range and Cascade foothills. The newcomers needed wood for their houses and barns and fences. Some of the earliest entrepreneurial efforts in the new land were devoted to sawmills. Elias Briggs built Springfield's first sawmill in 1853.[4]

Courtesy of Lane County Historical Museum

The Springfield Roller Mills, c. 1905

The gigantic trees up the river canyons were mostly of no interest to the early fortune-seekers. These trees were immense—too big for the pioneer sawmills to handle even if some genie could be found to fell them and drag them down to flat land. It would remain for the second generation of pioneers and some brash newcomers to turn them into profit.

In 1895, the brothers Robert A. and Henry Booth and the brothers George and Tom Kelly started a lumber company in Springfield. Robert Booth, born near Yoncalla, in the Umpqua River drainage of southwestern Oregon, was

Courtesy of Lane County Historical Museum

A Booth-Kelly steam donkey, c. 1900

A Booth-Kelly steam locomotive, c. 1915

an experienced mill man, having built his first sawmill in 1882, cutting ties and timbers for the Oregon & California Railroad.

The new company had its sights on tens of thousands of acres of former O&C grant lands from the Southern Pacific Railroad, the O&C's successor.[5] In 1899 the partners built a new steam-powered electric[6] sawmill at Wendling, up the Mohawk River,[7] and another mill at Springfield. Within the next decade the town's population doubled to 700, and property values doubled or tripled.[8] Half the town's men worked for Booth-Kelly.[9] The company dominated Springfield's economy for more than half a century, surviving a devastating mill fire in 1911 and some shaky times during the Great Depression. Booth-Kelly drove Springfield's growth as a community and cemented its identity as a timber town.

Courtesy of Lane County Historical Museum

The Booth-Kelly mill at Wendling, c. 1920

Booth-Kelly's success was part of a region-wide boom in the timber economy around the turn of the twentieth century. Improvements in logging and milling technologies and the availability of pools of outside capital had made it both possible and profitable to log the Northwest's big trees, and the choice timber pickings attracted players both small and large. Some entrepreneurs, like Frederick Weyerhaeuser, D. A. Blodgett, and T. D. Stimson, came to the

Courtesy of Lane County Historical Museum

A Springfield couple takes the air in an early Cadillac, c. 1908

Pacific Coast from the exhausted pine forests of Minnesota, Michigan, and Wisconsin with deep pockets of capital, looking for new worlds to conquer.

In the decades surrounding the turn of the twentieth century, lumbering had become ruthlessly efficient, and, as a result, the industry had become more and more nomadic. The stripping of the Great Lakes region had been fast and brutal, reflecting the economics of the marketplace:

> Lumbering in the region of the Great Lakes became an industry of tremendous driving power. Its captains and kings were hardy and resourceful, skilled in organizing men and machines and in overcoming physical obstacles. Many of them were highly successful... Powerful economic forces were set in motion. Among them were speculation in timber, the amassing of huge properties, and a rapid increase in the capitalization of the industry.
>
> The drive of large capital investments for speed and profit brought about the rapid skinning of enormous areas of timberland. Twenty years, or even less, became the common lifetime of a sawmill. Then—dismantle, junk, and move on.[10]

Other operators were of the homegrown variety. Some of these also became big names regionally in the timber industry of western Oregon: Booth-Kelly, Bohemia, Giustina Brothers, Hult Lumber Company, C. D. Johnson, Willamette Valley Lumber Company. Together these companies cut a swath through the big woods of the western Cascades and Coast Range. But the trees were so huge and thick, and the country so vast and rugged, there was still plenty of timber left by the time the nation pulled itself out of the Great Depression.

New Pioneers

In the late 1930s, a new wave of pioneers arrived in Oregon. These were lumbermen from the South, winding up a three-decade run of logging the virgin shortleaf and longleaf pine forests of Arkansas, Louisiana, Mississippi, and southeastern Missouri.[11] One of them was Thomas Whitaker Rosborough, son of a captain in Robert E. Lee's Army of Northern Virginia. Rosborough was a commanding man, over six feet tall and weighing more than 200 pounds.

T.W. "Whit" Rosborough

> His appearance called for proper attention and respect....Close up, the man was even more formidable. Rosborough knew what he wanted and was not to be argued with. His employees could see it in his face, in the determined line of his mouth, his strong nose, ruddy complexion, and startling blue eyes. They could feel it in his voice—softly accented, southern, but a heavy voice, demanding, insistent. Plainspoken, blunt. And loud. [11]

T. W. Rosborough—everyone called him "Whit"—was coming to the end of a profitable 50-year career in lumber. His Caddo River Lumber Company had operated sawmills in the towns of Forester, Glenwood, and Rosboro, Arkansas, the last named after himself. He was seventy years old in 1938, and one might be forgiven for thinking he was ready to settle down into a peaceful retirement. But he was not the sort of man who slowed down gracefully:

> It is a fact that lots of water has gone over the dam since we last met and since the old Caddo was organized in 1906. We are all several years older, but I do not feel that I should quit, in fact, I am altogether too young to have such thoughts....
>
> I am seventy years of age; my father died at 76 and did not take anything like the care of himself that I do. My mother lived to be 93 and was killed in

Rosboro, Arkansas in modern times

Beuford S. Cole

> an automobile accident. I feel I have every reason to believe that I will live to be 80 years or older.[12]

Besides, he said, he had trusted associates who would carry on after he was gone. "One of these is Mr. B. S. Cole, whom we have slated as our Secretary and Treasurer."[13]

For Whit Rosborough, trust and loyalty were not only expected but demanded. In November of 1938, as he was preparing to move to Oregon, he wrote to Beuford S. Cole, his future general manager and right-hand man in the Rosboro operation:

> The more I think of my conversation with you this morning, the more I feel that no man should go west with us, unless he has faith in the success of the business. This I believe applies particularly to those employees in the upper brackets.
>
> The old saying, which I believe to be true, that a man that has no faith in his undertakings is licked before he starts would apply and we do not want to go west with a man that has this attitude or frame of mind.

> In a calm impassionate manner I suggest that you pass this word out to those who have funds that we have invited to invest with us, and I am sure that I am correct in my position on this subject. If I am willing to put my life's earnings of seventy years of age in this business, no man should hesitate to do likewise if he wishes to become an active member of the Oregon organization.
>
> Yours very truly, [signed] T. W. Rosborough[14]

Beuford S. Cole had served in the World War as a sergeant. After the war he went to work for several lumber companies, finally landing at Caddo River. By 1932 he was managing the Rosboro plant. Intelligent and industrious, Cole was a self-taught man who overcame his lack of formal schooling by educating himself through correspondence courses.

There is no record of the conversation mentioned in the first paragraph, but Beuford Cole clearly got the message. The following year, Cole would uproot his wife, Teden, and their three young children, Ruth, Paul, and Elizabeth, from a comfortable middle-class existence in Rosboro, Arkansas, and follow his imperious employer out west. Cole would become later general manager, and eventually partner, in the Rosboro Lumber Company, and his son Paul would follow in his footsteps.

Caddo River

Back in Arkansas, Rosborough had been a prominent and successful lumberman since 1906, when he and three partners organized the Caddo River Lumber Company. M. R. Smith had the largest investment in the new company; W. E. Cooper, a lumber salesman, was enlisted to sell the new company's products; and Lee Wilson's job was to keep the books. Rosborough, the second-largest shareholder, had the hands-on responsibilities of building, equipping, and operating the company's first sawmill. Caddo River would grow to become one of a dozen big companies and hundreds of smaller ones that would log out the pine forests of the Ouachitas, the largest shortleaf pine forest in the world, in the first one-third of the twentieth century.

Rosborough, born three years after the end of the Civil War, was brought up on a thousand-acre plantation near Texarkana, Texas, by his Confederate-officer father and his genteel Yankee mother. The family also had a town house in Texarkana. The family's fortunes evidently survived the turbulence of Reconstruction, for young Thomas Whitaker Rosborough and his seven sisters had a quietly well-heeled postbellum upbringing. Young Whit was

Whit Rosborough in his school uniform

sent north to be educated at a military school near Chicago and returned to finish his schooling at an Episcopal preparatory academy operated by the University of the South in Sewanee, Tennessee.

At an early age the boy exhibited the restless impatience that would dominate his dealings as an entrepreneur, manager, and businessman. His grades were tolerable; he was pretty good in mathematics, but after a year at the prep school and a semester at the university he dropped out. "…during the last grading period at Sewanee, he had collected an unusually large number of demerits for conduct, so it appears he was simply fed up with school." He came home and went to work for his father, in the elder Rosborough's Bowie Lumber Company.

After two years there he quit and bought his own sawmill for $800. He was twenty-one years old. He ran the T. W. Rosborough Lumber Company for four years, then sold it and bought a bigger one in Louisiana. But his timing was inopportune. The Panic of 1893 hit hard, and lumber prices fell. Young Rosborough lost the mill and embarked on a decade of learning by doing. After paying off his creditors, he worked for a few years selling lumber in Missouri and Kansas. He took a job running a planing mill in Havana,

Young Whit

Arkansas, boarding with the town's doctor, John Mitchell. Mitchell had a pretty twelve-year-old daughter, Urcy, whose life would later intersect with Rosborough's, causing him much joy and much pain.

Eventually Rosborough became a partner in a lumber wholesaling business. Around 1905, in his mid-thirties and still unmarried, he moved to Prescott, Arkansas, where his sister Elizabeth lived. Elizabeth Rosborough was married to William N. Bemis, whose family had been in the lumber business for a hundred years, first in the Northeast and then in the South. Bemis's father was a partner in the Ozan Lumber Company in Prescott. Rosborough may have worked for Ozan, and he certainly found good friends and future supporters in members of the Bemis family.

Rosborough had gotten back on his feet and he was ready to get into sawmilling again. He bought two farms in Pike County, western Arkansas, along the Gurdon & Fort Smith Railroad, paying their owners $10 an acre. He built a town on the site, and he named it Rosboro—simplifying the spelling of his name "for trade and postal purposes."

For the next couple of years Rosborough went looking for timber. He bought about 2,400 acres of land and timber rights from his sister's in-laws,

Whit Rosborough in a formal portrait

the Bemises, paying about $1.35 per thousand board feet. From a neighboring lumber company he bought 3,500 acres and cutting rights on another 1,000 acres. The timber lay in western Pike County, along the Little Missouri River.

By 1907 the sawmill was complete. Rosborough employed both white and African-American men to work at his mill. The black workers came to town and settled their families in a tent ghetto while they looked for more permanent housing. Their presence caused some hostility among the whites, and Rosborough responded by building a high board fence around the tent community to protect his employees. Eventually they and their families moved into houses in the town's African American neighborhood, which was called, typically for the time, the Colored Quarters, or, more plainly, Nigger Town.

Relations between the races were generally peaceful but with an uneasy undercurrent. Rosborough was a man of his time, and he never challenged the prevailing racial segregation of the day, but he would willingly hire blacks to work in his mill, and some whites resented this. For the black employees, it was a good deal. Mill work paid better than field labor or domestic work, then the primary ways for African Americans to make a living. The black men in Rosborough's mill, as in other mills, tended to be assigned the heaviest jobs, such as stacking green lumber. Sometimes white men worked alongside them, and in that job, at least, men of both races received the same pay. Rosborough was no pioneer of racial integration, but it was generally acknowledged that he treated his employees, both black and white, fairly.

"Mr. Rosborough always liked the black people," recalled Avenelle Smith, whose father, Beriah Hutchings, made moldings in the Rosboro mill. "He thought they were a good group. We all liked them."[15]

In 1906 Rosborough had married Winifred Melville, an Englishwoman whom he probably met in Prescott, where Elizabeth and William Bemis lived. The wedding took place in Arkadelphia, a town along the Ouachita River about forty miles from Prescott, on the Missouri Pacific rail line that led to Little Rock. Then Rosborough took his bride back to Rosboro, where he had established a small mill to cut timbers for the big one he was planning to build. He built his wife a spacious bungalow with 10-foot ceilings, running water, indoor toilets, and a telephone. It was the fanciest house in town. Their son, Thomas Whitaker Rosborough, Jr., was born in 1907, and soon afterward the couple moved into their new home, from which they could see the sawmill going up.

After the mill was completed, the company began to sell lumber and make money. Rosborough made his daily rounds at the mill, and Winifred took care of the house and her son and tended to charitable works in town. Rosboro was a company town, with log trains rumbling through every day. The Missouri Pacific Railroad bisected the town, separating Rosboro's two hundred fifty black residents from its two hundred fifty white residents. Each side had its own church and general store. The church on the white side of town was served variously by Methodist, Baptist, and other Christian pastors. The roads were gravel, the sidewalks were board. In the sleepy Southern summers, the town's children could walk the mile to the Caddo River to go swimming.

In 1912, when little Whit was four, Winifred became pregnant again. This time, however, the pregnancy ended in tragedy. She miscarried, hemorrhaged, and died. She was thirty-four years old.

Another of Rosborough's sisters, Nelle Beidleman, came from her home in Prescott to stay with her brother and help ease him through this upheaval in his life. The most urgent question was, who would care for the little boy? Nelle Beidleman went back to Prescott and returned with a teen-aged African American girl named Halcyon.[16] Halcy, as she was called thereafter, proved to be a loving nanny, an efficient housekeeper, and a good cook. She soon took over the management of the whole Rosborough household. Halcy would grow up to marry Fred Wingfield, whom Rosborough later hired to be his chauffeur. The Wingfields would stay with Rosborough for the rest of their lives.

Fred and Halcy Wingfield

Nelle Beidleman had twin daughters, Mary and Martha, and a third daughter, Elizabeth. Martha was a particular favorite of Whit Rosborough. Her husband, Tom Watts, had an unmarried sister, Anna. Tom, Martha, and Anna would all come to play major roles in Rosborough's new venture in Oregon.

Caddo River Lumber Company continued to prosper. A 1915 fire led Rosborough to rebuild and modernize the Rosboro mill. By then he was down to his last seven years' supply of timber. With the help of L. D. Williams, the log hauler and railroad surveyor who had become his logging superintendent, he bought more virgin timber in neighboring Montgomery County, deeper into the wooded ridges of the Ouachita Mountains.

In 1916, after four years a widower, he remarried. His new wife was Urcy Mitchell, the daughter of the family with whom he had boarded while operating the planing mill at Havana, Arkansas. Urcy was twenty-nine; Whit was forty-eight. The newlyweds took a honeymoon trip to the Pacific Northwest.

It was the first time Rosborough had seen the forests there, and he was stunned by their rugged magnificence. He learned that a man could get almost a quarter-million board feet to the acre from those stands of virgin Douglas fir. Arkansas shortleaf pine forests yielded perhaps 5,000 board feet to the acre. Rosborough was as impressed by the mathematics as he was by the majesty of the forests—perhaps more.

In 1922, he expanded his Caddo River operation when A. L. Clark Lumber Company, in nearby Glenwood, cut the last of its timber.

> Instead of acquiring more raw material in Arkansas, A. L. Clark… had purchased timber on the West Coast. A man given to flowery oratory, he declared in a speech to fellow lumbermen that the South was his mother but that the great Pacific Northwest was his wife, and so he would move to Oregon.

The purchase would make the Caddo River Lumber Company the second-biggest in the Ouachita Mountain region of west-central Arkansas. But personal difficulties were making life hard for Whit Rosborough. Urcy was an attractive woman. Rosborough suspected an affair, and his suspicions proved true. When he confronted her with the evidence—a hired detective had seen her and her lover together in a hotel room—he confronted her and dismissed her as summarily as he might an unsatisfactory employee. He gave her a few days to pack her things and leave. Then he divorced her, complaining bitterly to Beuford Cole that he had, after all, given Urcy a high social position, and this was how she thanked him!

After the divorce Rita Tuppin, sister to Winifred Rosborough, came to keep house for Whit Rosborough. "We were entertained many times in this house," Paul Cole recalled, "and Mrs. Tuppin was indeed a gracious and friendly person. She also got along very well with Mr. Rosborough. I do not know if there was any romance there, but when Mrs. Tuppin later died, my father told me TWR took it very hard."

In 1929, two years after the divorce, Rosborough started a third sawmill 30 miles north of Rosboro in Scott County, on a site along the Fourche River. There was plenty of timber, but the site was a long way from the railroad. A prominent Scott County businessman, Charlie Forrester, persuaded the Kansas City Southern to extend a line into the area. Rosborough named his new mill and town Forester, both to honor his associate and to evoke the forested setting of the place—hence the dropping of the second "r".

By July of 1931, the mill was in operation, and two years later T. W. Rosborough, Jr., who had just graduated from MIT with an engineering degree, became its manager. Young Rosborough, nicknamed Bill,[17] had had a hard time getting through college, but when he came home to work for his father, he was all business, full of innovative ideas, nervous and intense, a fast walker and perhaps a fast driver, too.

Caddo River had struggled through the first two years of the Depression, but unlike many of its counterparts had managed to keep both the Rosboro and Glenwood mills open by cutting wages and running short shifts. The building of the Forester mill had taken most of the company's operating capital, and now, with orders plummeting, the managers feared they couldn't keep it going. In the midst of these financial worries, Rosborough suffered a third personal tragedy in 1934 when young Whit, driving toward Arkadephia, hit another car head-on. The crash broke his neck, and he died that night in a Hot Springs hospital. He was twenty-six years old.

By 1938, the Caddo River Lumber Company had been operating for more than three decades. Its owners were getting on in years, and the timber was nearly gone. The idea of milling second-growth timber was not much on anyone's mind. Timber was widely regarded in those days as a one-time windfall, something to be mined until it was exhausted. In just a couple of years Weyerhaeuser would pioneer the concept of a "tree farm," the notion of growing timber as a farmer would grow corn, to be planted and harvested in perpetuity. Indeed, the very term "tree farm" would be coined in 1940 by a Weyerhaeuser public-relations man, Roderic Olzendam, to name the company's new conifer plantation near Montesano, Washington.[18] But not many operators had such forward-looking ideas.

After his son's death, the elder Rosborough, then in his sixties, buried his grief in work. He continued to manage the company through the 1930s. Throughout that decade, Caddo River, along with many other companies in the region, had been selling off its cutover land to the Forest Service, which then added it to the acreage of the Ouachita National Forest. In the depths of the Great Depression, stumpland was not much in demand, and the Forest Service was paying prices in line with the land values of the day, as little as $1.10 an acre. In 1934 Caddo River's owners sold off 84,000 acres of timberland to the Forest Service for $1.25 an acre, retaining cutting rights to the old-growth timber that the company had planned to feed the 2-year-old

Forester mill.[19] The following year they sold off another 69,000 acres for $2 an acre. There were more land sales in 1938 and 1939, and in the end the company sold the Forest Service about 190,000 acres.

Winding Things Down

Caddo River's Glenwood mill shut down in 1938. In January of the following year, the Rosboro mill sawed its last log. Although the Forester mill would continue to operate until 1952, Caddo River's shareholders began liquidating some of the company's assets. All during this time, Whit Rosborough had been quietly making plans for his future business enterprises, and those plans lay far to the west.

In 1937 he bought a tract of timberland in Lane County, Oregon, a huge county stretching from the Pacific Coast to the crest of the Cascades. Finally he was acquiring some of the magnificent Douglas-fir forests that had so impressed him on his honeymoon trip twenty years earlier. There is no record of his thoughts as he made the decision to buy. Perhaps the experience was more bitter than sweet, recalling memories of his failed marriage and his dead son.

Nevertheless, from a business standpoint it was an excellent bargain. Returning from a family vacation-cum-scouting expedition to the West Coast in 1937, Rosborough's trusted assistant, Beuford Cole, reported that:

> I have just returned from Oregon, and the proposition looks good to me, and I am going out there with Mr. Rosborough when we start this new company. I found his timber of excellent quality, old growth Douglas fir, and found mills near him were paying $1.75 per M' for such stumpage. He has his timber bought at a cost of less than .50¢ per M'...[20]

Rosborough had initially purchased 343 million board feet of virgin Douglas-fir timber on about 10,700 acres in the Coast Range of western Lane County[21] near Mapleton. At less than 50 cents per thousand board feet, the timberland cost him something under $171,500. The tract was near the Southern Pacific railroad line and the Siuslaw River, and Rosborough planned to build a single-band mill at Mapleton that could saw 100,000 board feet per 8-hour shift.[22] In 1938, however, he found a timber deal more to his liking: about 20,000 acres of timber along Quartz Creek in the drainage of the McKenzie River. The timber was of better quality than the

Mapleton timber. It lay in the Cascade Range foothills of Lane County, in a single township adjacent to the McKenzie River Highway, a paved blacktop road. It stood thirty-five miles from Springfield, a town of two thousand five hundred people at the confluence of the McKenzie and the Willamette rivers. Rosborough changed his plans—he would build his mill in Springfield. The hauling of the logs would be literally downhill all the way.

The McKenzie River, looking upstream from Finn Rock

The timber was owned by the McKenzie River Timber Company. Rosborough negotiated a trade: his Mapleton timberland plus $50,000 for nearly a whole township up Quartz Creek. Rosborough walked away with 700 million board feet of old-growth Douglas-fir and hemlock for 35 cents a thousand board feet.[23]

It was a good deal, if not the unbelievable bargain it may seem today. "As a matter of historical perspective," said Paul Cole, "even after I began work at Rosboro Lumber Company in the 1950s, we could have purchased timber only 10 miles from the plant for less than one dollar per thousand board feet. The problem was where to get the one dollar."

T. W. Rosborough was well pleased with the transaction. "The first tract only ran about 35 M' per acre, while the tract now owned runs 75 M' per acre, and is a better quality of timber, cheaper to log and worth more per M' than the first tract," he wrote to a prospective investor in the new venture.[24] Timber in nearby national forests was selling for $1.50 per thousand, and in Washington state, timber of similar value was going for $2.50 per thousand.[25]

The next year, 1939, Rosborough bought another 40 or 50 million board feet.[26]

The McKenzie Highway

Financing the New Venture

Meanwhile, through 1938 and 1939, Rosborough and Beuford Cole were sending letters to prospective stockholders, urging them to invest in the new Rosboro Lumber Company, as it had been named.

> Dear Mr. Luther, I am writing you at Mr. Rosborough's request. Mr. Dill or someone there told him that you wanted some stock in his new Oregon mill and timber proposition. We are not ready to issue stock now, but would like to know the amount you would like to subscribe for...[27]

> Dear Mr. Ponder, If you plan to go West with us, we thought you would perhaps like to have some stock in the company.[28]

> Dear Mr. Whitted: I am writing you at Mr. Rosborough's request to see if you would be interested in taking some stock in the new proposed Oregon

proposition. We do not want any money now, but would like to know the amount of stock you would like to subscribe for...[29]

Dear Mr. Dismukes: As you perhaps know, Mr. Rosborough is planning to organize a new lumber company to operate in Oregon, and I am to go out there with him. We thought you might like to make an investment in this company...[30]

In keeping with Cole's sober, careful outlook, his letters describe the new company's prospects in mutedly optimistic tones. The timber had been a good bargain, and "[b]ased on the past history of the lumber industry," wrote Cole to a prospective investor, "the value of the original or virgin stands of timber will increase as such timber is cut and becomes scarcer, and the last great stands are in the Northwest."[31]

Between them Cole and Rosborough cast a wide net for supporters of the new venture. They approached shareholders in the Caddo River Lumber Company, urging them to buy Rosboro shares with their Caddo River stock. They approached former suppliers of the Caddo River mills, lumber wholesalers who had sold the company's products, bankers and accountants who had worked with Caddo River, company employees, and other associates, friends, and family members.

Their goal was ambitious: to raise $500,000 by issuing ten thousand shares at $50 a share. Based on the value of the timber asset, the stock, Cole pointed out to prospects, was really worth $100.[32] For some, Cole sweetened the pot with the promise of a job in the new company. To R. E. Dismukes, who had been an employee of the company at the Forester mill,[33] Cole wrote:

You spoke of taking some stock with us when you were at Forester. Now that you have left there, will you let me know if you are still interested in taking stock and going out there with us. If you would like to go out there with the new company and have a job with them, we would like for you to make your subscription as large as possible, as our going depends on whether we can sell sufficient stock and properly finance the operation.[34]

Then, in a bit of subtle arm-twisting, he reminded Mr. Dismukes of their employer's expectation of unquestioned loyalty:

Mr. R. is putting about everything he has in the proposition, and he feels that men who are interested in going out there should have sufficient faith in the proposition to make a substantial investment too in accordance with their means.[35]

Six months later Dismukes told Cole he would invest $3,000. While he wasn't trying to buy a job, he hastened to say, he really wasn't in a position to put up that much money without some prospect of employment.[36] Rosborough's response was testy. "I am glad you made no commitment to Mr. Dismukes," he wrote to Cole,

If he wants to invest $3,000.00 with us without pledging him a job, will be glad to receive his stock subscription, but if he does not want to invest with us without our pledging him a job,...we will not need his money.

In his cultivation of long-time colleagues and friends, Cole sometimes revealed a hint of sentimentality. To W. E. Cooper, an old acquaintance who had been one of Caddo River's original investors, he wrote:

"We too are saddened by the passing of the Rosboro and Glenwood mills and the beautiful stands of soft pine that once adorned these fair uplands of Arkansas, and shall ever remember the good old Caddo and what changes have come to us during the years. I hope that Rosboro Lumber Company may be equally successful and that time and fate may deal kindly with us all."[37]

Not everyone got on board. There were more than a few polite demurrals, like this one, from a Hot Springs physician with safer places to put his money:

However, I would not have any money at this time to invest. As you know, I have a little subdivision here that I am playing with for diversion, etc., and am planning on building some houses in this, which will use whatever cash I have on hand. Although, as you say, I am quite sure that an investment with the lumber company would be an excellent one over a long pull."[38]

Or this, from another doctor in the same building:

I regret that I will not be able to make an investment in the company but at this time most of my surplus is invested in annuities and in stocks that are not readily marketable.[39]

As Cole was soliciting financing from private investors, he was also negotiating for a half-million-dollar loan from the Reconstruction Finance Corporation. The RFC was a federal program established during the Hoover administration to restore confidence in America's Depression-ravaged banking system. The program had been expanded during Franklin Roosevelt's New Deal to provide government loans to assist, among others, agriculture, housing, exports, and business.[40]

A memo from Cole to Rosborough about the loan application outlines the company's plans:

Amount of loan requested,	$500,000.00	
To be used for:		
Plant Buildings & Machinery,	$438,493.00	87.7%
Logging & Road Building Machinery,	40,000.00	8.0%
McKenzie River Bridge,	15,000.00	3.0%
Miscellaneous (Overhead during Const)	6,500.00	1.3%

For collateral, the company was putting up the future mill, which would be worth $438,493 when it was completed, along with the logging and road building equipment and the planned McKenzie River bridge, together worth $55,000. The key asset, of course, was the timberland, appraised at $996,742.50.[41] The figure after the decimal point suggests that Cole was a careful and precise man.

The memo also lists some of the people who pledged support to the company at a May 1938 meeting:

To employees & others who subscribed at the meeting at Mr. Rosborough's office, approximately,	$25,000.
Mr. Cooper,	10,000.
Mr. Rhodes,	10,000.
Jess Hollis,	5,000.
Mrs. Beidleman & Mr. Watts,	20,000.
Other relatives of Mr. R & Mrs. Baker,	12,000.
	82,000.

W. E. Cooper, as we have noted, was one of the original investors in the Caddo River Lumber Company. Mrs. Beidleman was Rosborough's sister Nelle, who had come to his aid after the tragic death of his wife, Winifred, and who had hired Halcy Wingfield as a nursemaid for young Whit Rosborough. Tom Watts was married to Rosborough's favorite niece, Martha.

As of January 1940, shortly after its incorporation, Rosboro Lumber Company had twenty-six shareholders, including T. W. Rosborough, Beuford and Teden Cole, and R. T. (Tom) and Martha Watts.[42] Its first directors were shareholders Rosborough, Cole, George Bracher, Andreas Bracher, and L. V. Lundberg. The Brachers, of Portland, were dealers in timberlands and sawmill properties. Lundberg was a Portland forest-industry consultant.

In the very early days of the company George Bracher had evidently asked Cole for the first check written on the new Rosboro account, for sentimental reasons. Accordingly, Cole sent him check no. 1, made out to Bracher, in the amount of $2.[43] Later Bracher returned the check to Rosborough "which I was keeping for sentimental reasons" because "you may care to have it for the same reasons."[44]

Rosborough held the lion's share of the stock: a total of three thousand five hundred shares, or $350,000 worth, to cover his purchase of the McKenzie River timberland, as well as $100,000 worth of stock from Caddo River that he was contributing to the new company.[45]

The RFC loan was approved in January of 1939, at 5 percent interest.[46] A restrained excitement shines through Beuford Cole's words:

> We now know definitely that we are going ahead with the formation of the company. I went to Washington, D. C. with Mr. Rosborough on the 7th of this month, and we arranged for a loan of $500,000 from the RFC… We are highly pleased with this commitment, and while there I knew of one lumber operator being turned down. We were told that not in their history had a loan for this amount been approved by them in so short period of time. This, we think was due to our proposition having merit; our reputation as operators and having friends in high places at Washington who were most kind in helping us,…[47]

From Cole's copious correspondence—most of it typed with his own hand—emerges a picture of a man who was sober, careful, fussy, determined, and often long-suffering. He surely must have felt bullied occasionally by his strong-willed boss. But he'd been with Rosborough for a long time, and the men were closer than most business colleagues, closer even than most friends. Cole's son Paul had been born in Rosborough's big house in Rosboro. No one knew better than Cole the old man's quirks and tempers. Cole's hardworking habits may have been driven by a bit of insecurity. Perhaps he was uncertain about his prospects should he leave Rosborough and strike out on his own.[48]

There is no question that he displayed the loyalty his employer required. "I would not want to guarantee anything," he wrote to an initially reluctant investor, Dora Baker, "but if you wish to invest...will be glad to sell you the stock,...but if you do not feel safe in making the investment, we do not want you to make it." Nevertheless, he wrote, "I own more than $300,000.00 worth of the capital stock, ... My life's earnings are invested in the Rosboro Lumber Company." His words must have inspired Mrs. Baker's confidence, for two days later she invested $4,000 and received 80 shares of Rosboro stock.

Moving West

In March of 1939, Whit Rosborough was ready. He was going to Oregon, and he was taking thirty-nine of his Arkansawyers and their families with him, including two who would become prominent in the new company's leadership:

> My assistants will be B. S. Cole and Tom Watts. Mr. Cole, a most efficient office executive, has been my first assistant in the Caddo operation for a number of years. Mr. Watts, who married my niece, I think will develop into quite a good man under my training, and has sales experience.[49]

In late March of 1939, Rosborough threw a farewell party for his many friends and associates. A few days later, on the twenty-fifth, he climbed into the backseat of his big Chrysler. Fred Wingfield was at the wheel, with Halcy beside him. Rosboro directed his chauffeur to drive not west, but east, toward Huntsville, Alabama. He had some personal business to take care of before the long trip.

Beuford Cole knew something was up. "I only heard about it a few days before he left," he wrote to W. E. Cooper, "and began to think he wasn't going to tell me about it before leaving, but he finally told me the 'secret' before getting away."[50]

The secret was that, for the third time, at seventy years of age, Rosborough was getting married. His bride was Anna Watts, the forty-eight-year-old sister of Tom Watts. Martha Watts had introduced Whit to Anna years before. Somehow she managed to get her uncle and her schoolteacher sister-in-law together the previous winter, and he proposed marriage.[51] "I hope he may be very happy," wrote Cole.[52]

Beuford Cole and his family stayed behind in Rosboro until May, while he wound up the business of closing Caddo River's books and shipping mill machinery to Springfield.[53] Besides, they wanted to wait until the children were finished with the school year. Teden Cole was understandably ambivalent about the move. Rosboro was a small town, certainly, but it provided a comfortable middle-class existence, with church and community and wholesome activities for the children. Ruth was already finished with high school and ready to go on to college.[54] Paul was ready to enter junior high school.

Being brought up in Rosboro had given the children what today would be termed an idyllic childhood. There were chores at home, cows to milk and chickens to tend, firewood to split and coal to haul, and the Caddo River to cool off in on hot summer days.

"I remember Paul when he was six, or maybe not that old," said Avenelle Smith, now 95, whose family, the Hutchingses, lived next door to the Coles in Rosboro. She was in her late teens then. "Mrs. Cole used to bring the girls [Ruth and Elizabeth] over and visit." The Cole girls would always go for the family's pile of magazines, spreading them all over the floor and playing with them. "Mrs. Cole would laugh about it. She'd say, 'Avenelle outgrew those things, and they will too.'"

The families were not social equals, Avenelle Smith said—"they were our employers, even though they lived right next door. But we always thought a lot of them."

Teden Cole had ambitions for her children. She urged them to pay attention to their schooling, pushed them to go on to college, insisted on proper table manners.[55] The prospect of moving to Springfield, Oregon may have seemed to her something like a sojourn in the wilderness. How would she keep up a proper household in such remote and rugged surroundings?[56]

If so, she was not the only wife with misgivings. R. E. Dismukes, who had inquired about employment in exchange for his investment, was in fact offered a job in Springfield, but in August of 1939 he sent a telegram to Beuford Cole saying he was changing his mind and staying in Arkansas. The next day, August 24, came another telegram: "Disregard my telegram yesterday and letter acted hastily ... will write you fully explaining my hasty action yesterday."[57] In a letter dated the same day, Dismukes offered this explanation:

> Don't judge me too harshly about the telegram and Air Mail letter of 23rd. Just a family row I had with my wife's brother and her sister about her moving west, and I got mad and my wife cried, and I acted on the spur of the moment.[58]

Mrs. Dismukes, however, evidently came around: "Mary likes Oregon and she is with me in anything I do."[59] Cole's reply must have reassured him: "Your letter of the 24th was received this morning, and clears things up, and I can understand how this might have come about."[60]

By July of 1939, Rosborough and his principals had settled in Springfield after a cross-country drive from Arkansas, with Fred Wingfield at the wheel of Rosborough's Chrysler. Mill machinery was arriving every day. The bridge over the McKenzie River, which would allow the company to get to its township's worth of timber, was under construction. The mill was rising east of town.

The company opened an office on the second floor of Fifth and Main, above Gerlach's drugstore and soda fountain. The arrival of a passel of Arkansawyers, and a big man in a big car with a black man for a driver, and the prospect of a new sawmill in town, had Springfield all abuzz.

Notes

1. Letter from T. W. Rosborough to B. S. Cole.
2. Velasco, Dorothy, and Mara Velasco. *Springfield: Between Two Rivers* (Montgomery, AL: Community Communications, Inc., 1999), p. 20.
3. Boyd, Robert. *Indians, Fire, and the Land in the Pacific Northwest* (Corvallis, OR: Oregon State University Press, 1999).
4. McChesney, Jim. "Timber: a challenge then and a challenge now" (Springfield, OR: The Springfield News, Centennial Edition, 1985).
5. Beckham, Dow. *Swift Flows the River: Log Driving in Oregon* (Coos Bay, OR: Arago Books, 1990), pp. 39-40.
6. Springfield Historical Museum display, visited 7/7/05.
7. Ibid., Velasco and Velasco, p. 32
8. Ibid., McChesney.

9. Ibid., Springfield Historical Museum.
10. Greeley, William B. *Forests and Men* (Garden City, NY: Doubleday, 1951), pp. 40-41 cited in Smith, Kenneth L,. *Sawmill: The Story of Cutting the Last Great Virgin Forest East of the Rockies* (Fayetteville, AR: University of Arkansas Press, 1986), pp. 3-4.
11. Smith, Kenneth L. *Sawmill: The Story of Cutting the Last Great Virgin Forest East of the Rockies* (Fayetteville, AR: University of Arkansas Press, 1986), p. 4. This book is the source of most of the material about Rosborough's early life in Arkansas.
12. Letter from T. W. Rosborough to Milo T. Jones, Feb. 15, 1939.
13. Ibid., Rosborough to Jones.
14. Letter from T. W. Rosborough to B. S. Cole, Nov. 17, 1938.
15. Interview with Avenelle Smith, Nov. 30, 2006.
16. The adjective "halcyon," meaning happy, peaceful, or idyllic, comes from the name of the Greek demi-goddess Alcyone, who represented tranquility.
17. Paul Cole said, "All of the former Caddo people I talked with referred to TWR junior as 'Bill,' who was apparently very well liked by them. Have no idea where the Bill name came from however."
18. Interview with Clarence Richen, June 9, 1994.
19. The Forester mill, the last of Caddo River's three, would continue to operate until 1952.
20. Letter from B. S. Cole to W. T. Luther, July 22, 1937.
21. Letter from T. W Rosborough to Ralph M. Rounds, Feb. 14, 1939.
22. Letter from B. S. Cole to C. E. Forrester, Aug. 3, 1937.
23. Letter from B. S. Cole to R. E. Dismukes, Nov. 17, 1938.
24. Ibid., Rosborough to Rounds, Feb. 14, 1939.
25. Letter from B. S. Cole to Mrs. D. H. Baker & Daughter, Jan. 30, 1939.
26. Ibid., Rosborough to Rounds, Feb. 14, 1939.
27. Letter from B. S. Cole to W. T. Luther, July 22, 2937.
28. Letter from B. S. Cole to Clarke Ponder, July 26, 1937.
29. Letter from B. S. Cole to Alph Whitted, July 24, 1937.
30. Letter from B. S. Cole to R. E. Dismukes, Aug. 3, 1937.
31. Letter from B. S. Cole to J. A. Borland, Jan. 30, 1939.
32. Prospectus for Rosboro Lumber Company, Jan. 30, 1939.
33. Letter from B. S. Cole to T. W. Rosborough, May 22, 1939.
34. Letter from B. S. Cole to R. E. Dismukes, Nov. 17, 1938.
35. Ibid., Cole to Dismukes, Nov. 17, 1938.
36. Letter from B. S. Cole to T. W. Rosborough, May 22, 1939.

37. Letter from B. S. Cole to W. E. Cooper, March 20, 1939.
38. Letter from A. H. Tribble, M.D., to B. S. Cole, May 16, 1939.
39. Letter from Francis J. Scully, M. D., to B. S. Cole, April 4, 1939.
40. The EH Net Encyclopedia, http://www.eh.net/encyclopedia/?article=butkiewicz.finance.corp.reconstruction, consulted Sept. 10, 2005.
41. Memo from B. S. Cole to T. W. Rosborough, regarding RFC loan application. The memo is undated but mentions a meeting of prospective stock buyers at Rosborough's office in Glenwood, Ark., held May 21, 1938.
42. Minutes of stockholders' meeting, Rosboro Lumber Co. January 2, 1940.
43. Letter from B. S. Cole to George M. Bracher, July 25, 1939.
44. Letter from George M. Bracher to T. W. Rosborough, Nov. 12, 1943.
45. Ibid., Bracher to Rosborough, Nov. 12, 1943.
46. Prospectus for Rosboro Lumber Company, Jan. 30, 1939.
47. Letter from B. S. Cole to C. W. Rabb, Jan. 28, 1939.
48. Interview with Ruth Cole Kainen, Aug. 10, 2005.
49. Letter from T. W. Rosborough to Ralph M. Rounds, Feb. 14, 1939.
50. Letter from B. S. Cole to W. E. Cooper, April 18, 1939.
51. Or maybe she proposed to him, suggested Ruth Cole Kainen. "Just before he left Arkansas, Anna fortuitously came to see him," she said in a 2005 interview. "She was a direct woman."
52. Ibid., Cole to Cooper, April 18, 1939.
53. Letter from B. S. Cole to D. L. Johnson, May 24, 1939.
54. Ruth Cole went on to major in English at the University of Oregon and excelled. After college she joined the U.S. Naval Reserve and became a chief petty officer, "a great accomplishment," said her brother Paul.
55. Ibid., Cole, Notes.
56. Interview with Ruth Cole Kainan, Aug. 10, 2005: "My mother told me later, 'My heart sank (when she saw Springfield). It's farther away from civilization than even Rosboro is!'"
57. Telegram from R. E. Dismukes to B. S. Cole, Aug. 24, 1939.
58. Letter from R. E. Dismukes to B. S. Cole, Aug. 24, 1939.
59. Ibid., Dismukes to Cole, Aug. 24, 1939.
60. Letter from B. S. Cole to R. E. Dismukes, Aug. 28, 1939.

2

A NEW START IN A NEW LAND

"Do not let anyone know anything about what wage scale is. If they do not have any more faith in my fairness than this, let them stay in Arkansas."[1]

–Thomas Whitaker Rosborough

THE SPRINGFIELD THAT the Rosborough entourage encountered in 1939 was a town of three thousand eight hundred people,[2] a working-class community supported mostly by lumbering and agriculture, a blue-collar cousin to Eugene, home of the University of Oregon, across the Willamette River to the west. People in Springfield were mill workers, store clerks, small-business owners, doctors and dentists, lawyers and teachers, farm workers. Around Springfield, encircling the smaller communities of Mohawk, Wendling, Thurston, and Jasper, were farms growing hops, filberts, beans, corn, carrots, and beets.[3]

Hops had been a mainstay of Lane County's farm economy since 1882, when Alexander Seavey brought the first vines to Springfield.[4] Filberts (or hazelnuts, as today's producers would prefer to call them) were introduced into the area by Ben Dorris in 1935. Dorris established a nursery on his orchard, and for many years he produced about seventy thousand hazelnut seedlings a year.[5] The Dorris family was an important contributor to an industry that still thrives in Oregon today.

Lumber, though, was the bulwark of Springfield's economy and the signature of its way of life. The Booth-Kelly Lumber Company built its first large mill at Wendling, up the Mohawk River from Springfield, and in 1901 the company bought the sawmill that had been built by Elias Briggs and built a much larger mill and millpond at the base of Springfield Butte.[6]

The Booth-Kelly mill dominated the town, said Grady Lewis, a retired

Springfield High School teacher who grew up hearing the mill whistle blow the nine o'clock curfew. "When you heard that whistle," he said, "you knew it was time to go home."

Courtesy of Lane County Historical Museum

Young women hop pickers at Seavey hop yard, c. 1914

The Booth-Kelly sawmill was located on a long strip of land directly south of Main Street, bisected by a millrace. The mill was powered by a combination of steam power and electricity furnished by a large boiler plant, complete with turbines.[7] The long headrig was driven by a steam engine that made the characteristic *choo-choo* sound of steam locomotives. At times the engine would shake the windows in the downtown stores and offices. "Booth-Kelly seemed to like to have a lot of very noisy whistles blowing day and night, which really bothered their household neighbors," Paul Cole recalled. "Besides, there was a lot of very black smoke. You could tell when the sawmill was running by whether there was smoke coming out of the stacks." The Springfield Plywood plant to the east put out even blacker smoke, he said, from the very hot fires produced by the sander dust.

Like other towns across the region and the nation, Springfield was pulling itself out of prolonged hard times. The Great Depression had hit the Pacific Northwest's extractive industries hard. Ninety percent of Oregon's timber companies were driven to near-bankruptcy, and about half the state's private timberlands were tax-delinquent. [8] Everywhere, homes and businesses were foreclosed on and people were thrown out of work. In Springfield and the rest of the Willamette Valley, nature had added to the troubles when a plant disease called downy mildew devastated the hop yards and destroyed a once-thriving industry.[9]

Courtesy of Lane County Historical Museum

Booth-Kelly mill at Springfield, c. 1910

As the 1930s drew to a close, however, things were beginning to look brighter. The McKenzie River was a world-class fishing river, and a seasonal influx of eager anglers formed the beginnings of a recreation industry and made up the first customers for a newfangled, stubby, curvy-prowed, flat-bottomed wooden craft that has since gained a worldwide following: the McKenzie River drift boat, invented by Springfield boat builder Woodie Hindman.[10] In 1937, the town felt expansive enough to build a new movie theatre, the Playmore. The Springfield *News* praised the theatre's "large, comfortable seats" and "clear pictures on the large screen."[11]

Settling In

By July, Whit and Anna Rosborough, Beuford and Teden Cole, Martha and Tom Watts, and the rest of the Arkansawyers[12] and their families had settled into their new homes in Eugene, across the river from Springfield and the mill. "The painters and workmen are about through working at the office," wrote Cole to Nelle Beidleman, Rosborough's sister, "the bridge on the McKenzie is nearly completed, buildings are beginning to show up on the plant construction and we are really busy. We bought a home on the SW corner of 19th and Potter Sts., or just three blocks North of Mr. Rosborough, and think we shall enjoy the home very much when we get possession around August 1st."[13]

Anna Margaret Watts Rieder has warm memories of the Rosboroughs' home on Potter Street. Margaret Rieder, now 86, is the namesake of her aunt Anna, although she has been called Margaret since infancy (to avoid confusion between two Anna Wattses, she said).

Margaret came out from Huntsville, Alabama, in 1940 to go to school at the University of Oregon, and, at the invitation of her generous aunt and uncle, she stayed and boarded with them at the house on Potter Street. Her memories of her uncle Whit reveal a man warmer than his business dealings might indicate, a man with a rich and enjoyable private life. "He was the perfect gentleman," she recalled. "I never heard any cuss words, but he was a great tease."

For instance, when she'd get a phone call from a young man she was dating, her uncle would hand her the phone without comment. But after she hung up, she said, he'd be waiting with a grin on his face. "Did he say he loved you?" he would inquire meaningfully. "He was trying to make me blush," she said. "He could be so mischievous."

She remembered Whit and Anna playing cards in the evening with some of their many friends. At a certain point Whit would rise and say, with a smile, "It's time for all good honest people to go to bed"—the twinkle in his eye implying, she said, that those who stayed up late were less than good honest people.

Rosborough was, in fact, a man of regular habits, Rieder said, a diabetic who disciplined his food intake even though he loved to eat. She remembered his frequent remark, "Some people eat to live. I live to eat." Whenever Halcy Wingfield, a fine cook, prepared something especially delicious, he'd eat his dinner with great enjoyment, and then he'd say, "That was bully!"

Many things in Oregon were a pleasant surprise to Rosborough and his entourage, including the mild climate. "The weather is still nice," wrote Cole to J. L. Hollis, an equipment supplier he'd worked with back in Little Rock, in December of 1939. "Very little rain so far, which we are told is unusual for this time of year."[14]

Courtesy of Lane County Historical Museum

Sutton block at 5th and Main, Springfield, c. 1920

Pharmacist Claude Gerlach owned and operated Gerlach's Drugstore in the Arcade Building[15] at Fifth and Main, where the newcomers from Arkansas had taken upstairs office rooms. Gerlach, now in his nineties, recalled those days: "Things were pretty dead at that time, but the situation was getting better all the time. Booth-Kelly supported the town." Gerlach, who became Springfield's mayor in 1945, said Rosborough's Arkansawyers were mostly welcomed by the Springfielders. "We were excited to have the company here."

Rosborough's people seemed like one big family, said Gerlach. "They hung together and they were very faithful to him." He added, "Mr. Rosborough had a colored man for a chauffeur. That was quite a treat for the people of Springfield; there were no other blacks in town."

Louise Denning McQuin was hired by Beuford Cole in 1940 to take dictation and transcribe notes in the Rosboro office upstairs at Fifth and Main. Twenty years old and fresh out of business college, Louise Denning was living in Springfield and working at a Eugene wholesaler's office. One day a friend of hers who'd been sipping a Coke in Gerlach's soda fountain overheard someone say that the new company upstairs was looking for office help. She walked in, applied, and was hired.

At that time Whit Rosborough had a male secretary, a Mr. H. E. Davis, who also served as receptionist. The accountant was named Mr. Edom, McQuin recalled, and he stood at an old-fashioned high desk to do his work. "Mr. Rosborough had his office, and Mr. Cole had his," she said. "Where I was, there were two desks, just down the hall from [Cole and Rosborough's offices]. There was a large group of people in the other section." Initially her duties included dictation, typing, and calculating log volumes from scalers' notes. The vocabulary was all new and strange, with terms like "hog fuel" and "saw kerf" to be learned.

Her employers were soft-spoken and deeply courteous, and she found their Southern ways both strange and charming. It tickled her that they called her "Miss Louise." Comparing notes with her girlfriends who worked at other lumber companies—Booth-Kelly and, later, Weyerhaeuser—she said, "I felt sorry for them. They weren't treated like we were. We were spoiled." The Rosboro people "were always so respectful—they never talked down to you. Southerners don't raise their voices like we do."

She also recalled Al Carley, the company's hot-tempered sales manager, who was known to slam the telephone receiver down when he was frustrated, and sometimes to break wooden pencils in two. Carley was a small-town Oregonian, not an Arkansawyer, and he didn't get along with a lot of people, said McQuin, including the strong-minded woman who became Beuford Cole's (and later Paul Cole's) secretary, Lorene Burt. "She was a very capable woman, very well respected—a perfectionist, but with a sense of humor. She was funny, and she was sharp-minded. Mr. Cole called her Lorene. We called her Miss Burt."

Sometimes, Paul Cole remembered, he would slip up and call Miss Burt Leona, but she didn't mind too much, he said. "She would tell me 'Paul, I want to marry a sailor.' I would ask why, and her answer was, 'Because he would be gone most of the time.'"

Lorene Burt did all the secretarial tasks, plus log scale, plus other jobs, Cole recalled. "It didn't matter how much she was given to do—she was generally through by shortly after lunch time. You see, her real interest was to get time to do her crossword puzzles. I liked her very much and we became very close friends."

Louise McQuin has warm memories of Cole's father, Beuford Cole. "He was so very kind," she said. "If I made a mistake, he would correct me kindly." She found T. W. Rosborough slightly more formidable. "Being young, I was quite frightened of him. He was tall, large, and big-boned. You might have expected him to speak loud and roughly, but he was always kind to me. He probably saw me shaking."

Louise McQuin

She encountered more Southern hospitality when she made friends with a co-worker, Corinne Bridges, whose family had moved from Arkansas with Rosborough. "We'd walk to her house, and Mrs. Bridges would say, 'Louise, can you stay for lunch?' I was always available."

When she left in 1943 to enter the Women's Army Corps, Beuford Cole told her, "Miss Louise, you'll always have a job here." After the war Louise McQuin would come back to Rosboro and stay until she retired in 1987, except for five years away to care for her young children. McQuin worked in the sales department and eventually became the company's plywood salesperson.

Rosboro joined about a dozen other sawmills in the area, the big Booth-Kelly mill and a handful of smaller ones. Whit Rosborough and Beuford Cole set about acquiring equipment for the mill, which was rising rapidly on a 40-acre site Rosborough had bought from the city of Springfield east of town on the McKenzie River highway.[16]

One of the company's chief suppliers was Hollis & Company of Little Rock, Arkansas, which had furnished much of the equipment for the Caddo River Company's Rosboro and Glenwood mills back home. Hollis was the distributor for a dozen industrial-supply manufacturers, including Link-Belt Co., Simonds Saw & Steel, U.S. Rubber, and R. Hoe & Co. Its president, J. L. "Jess" Hollis, had a long-standing and warm relationship with Whit Rosborough and Beuford Cole, and he was one of the early investors in the new company.

Shed construction

Although Rosborough's new operation was far from Hollis's home base, he was keenly interested in keeping his old customers happy. In August of 1939 he sent a salesman, Joe Royer, out to Springfield to look over the Rosboro operation. "We did not plan the trip for Joe on the basis of the business Rosboro Lumber Company might favor him with," he hastened to clarify in a letter to Cole, but "we have for a number of years sent our men to various factories to gain what knowledge they can and to enjoy a vacation at the same time." Unquestionably, Hollis & Co. was interested in the Rosboro account, "but we do not want [Rosboro] to feel like we are trying to sell them any items other than those they would like to give us on an equal basis with others."[17]

Joe Royer, the salesman who made the trip, wrote to Cole in September to send along some photos he'd taken and to thank him for "the nice time shown me in Springfield." He praised the company's mill and, especially, its impressive tract of timber: "I am not a very good photographer but that timber was so thick, I had a hard time finding enough sunlight to make a picture." He thanked Cole for "the nice orders of saws and chain. They are the largest single orders ever sold in my territory."[18]

Later in the year, Jess Hollis suggested some equipment for Rosboro's planing mill from a manufacturer whose products Hollis & Co. carried. "The Jones and Orth company of Seattle, Washington, makers of Greyhound Heads, offer to work with us in trying to sell you Greyhound heads for your planing mill," he wrote to Rosborough and Cole in December of 1939, adding, "We do not want to seem too greedy in trying to sell you items from here. We appreciate from the depth of our hearts the business you have already given us." He went on to tout the quality of Jones and Orth's products and offer a free quote, "without any obligation on your part." He closed with, "We miss the nice orders we used to get from Caddo River at Rosboro and Glenwood we enjoyed for so many years, and miss the contact we had with you."

Planing mill

It was fine with Whit Rosborough to deal with old friends, but business was business; and, in fact, he disagreed with Hollis's estimation of the Jones and Orth cutter heads. "Mr. Rosborough is very partial to the Philbrick cutter head," Cole replied to Hollis, "and in fact says the Greyhound does not compare with the Philbrick." Diplomatically, Cole added, "This, of course, is a matter of opinion, but it is his opinion."

Rosborough also made it known to Hollis & Co. that he wanted them to split their profits on his orders with him.[19] "We might consider using the Greyhound," Cole wrote to Hollis, but "in this connection, Mr. Rosborough asked me to ask you if you were going to split the profit with us on this long range business." If not, the company could get discounts on some items, he told Hollis, by dealing with suppliers closer to Springfield.[20]

Planer

Cole put a bit of an apologetic spin on Rosborough's demand: "We are coming along very nicely with our plant, but it is going to cost more money than was planned.... With our cost over-running, we are trying to cut down every place possible."[21]

He smooths the waters further with his warm closing:

> We too miss the friendly associations of the past and the old days at Rosboro and Glenwood. I imagine Mr. Royer would have already been down to go bird hunting with us if we were still at Rosboro. However, we have to change in this changing world.[22]

Naturally reluctant to cut into his profits, Hollis made a tactful reply: "Now about Mr. Rosborough's question, we had not figured that he would want us to do any more than give him the same price he would have to pay the manufacturer direct, and as yet we do not know just what profit we are going to be allowed." However, "when the orders are shipped and we see just what profit we have in the sales, we will take the matter up with Mr. Rosborough and do what he thinks is fair."[23]

Cole's correspondence in the course of acquiring the mill equipment shows his customary caution and carefulness. He sought advice from other lumbermen on such details as the gauge of saw and the tooth spacing on the mill's 9-foot headrig,[24] and after inquiring of Sam Smith of Garibaldi, Oregon, the company's soon-to-be plant manager, about the specifications for the big edger, he got a second opinion.[25] The people of whom he made his queries were free with their advice to a potential competitor.[26] Such collegiality among rivals seemed to be common in those years in the jostling brotherhood of the Pacific Northwest lumber industry, among the established companies and the brash upstarts, where everyone knew everyone else, knew who had acquired timber and who was selling it, who had fired their head sawyer and who was looking for a woods foreman. Some of that friendly competition, or competitive friendliness, endures to this day.[27]

Green sorter

The bridge across the McKenzie was finished in late summer of 1939. The company threw a Sunday picnic at the bridge to celebrate.[28] By spring of 1940 the mill was almost ready to go—Cole expected it to "turn over" between the fifteenth and twentieth of March.[29] A laudatory article in a special edition of the *Springfield News* called it "one of the West's most modern sawmills,... While it is not one of the largest mills computed by the board foot cut on the head rig daily, the new machines and up-to-date design makes this new streamlined plant outstanding in the lumber industry."[30]

Sam Smith, the new manager, was evidently well versed in the requirements for saws to cut "the fur lumber,"[31] as he put it, for Cole queried him closely on the proper saw gauge and tooth space for the mill's headrig, resaw, big and small edger, and two trimmers.[32] Smith recommended 12-gauge saws with 3-inch tooth spacing for the head rig, 14-gauge saws with 2-inch tooth spacing for the resaw, 6-gauge saws with 24 teeth at 32-inch spacing for the big edger, and 9-gauge saws with 20 teeth at 20-inch spacing for the small edger "if the edger will take them."

Bridge over the Mckenzie at Finn Rock tree farm

The mill construction was chewing through money fast. In March Rosborough and Cole applied for another $100,000 from the federal Reconstruction Finance Corporation to finish construction and meet payrolls in March and April.[33] As it turned out, Cole's prediction of a mid-March opening date was optimistic. Still needed was about $88,000 worth of mill equipment, including an air compressor and other machinery and equipment from Allis-Chalmers, a cutoff saw from Stetson-Ross for the planing mill, some filing room equipment, some electrical equipment, and a sprinkler

Inside the sawmill, under contruction

system. The company also needed to pay off almost $23,000 on contracts for its dry kiln and lumber stackers and hoists.[34] Whit Rosborough, not a man for half measures, was spending more than he'd planned, which worried the careful Cole. "Mr. Rosborough is just building too much plant for the planned cost," he fretted.[35]

The RFC hired a San Francisco logging and sawmill consultant, Herbert Hessler, to check out Rosboro's situation. Hessler's report assured the RFC that the company was headed for success. He urged the government lenders to consider Rosboro's huge timber holdings and its location in an area where logs were plentiful and cheap. "...the demand for plywood peeler logs, at an approximate average price of $17.00 per M. will net this company an assured profit on its stock of this nature," he told them in his report, even if the mill did not make money right away. "It is my opinion," concluded Hessler, "that the operating set-up of this concern should cause it to be definitely successful." He predicted that the mill would start by the end of April.

Hessler's report presents the most succinct record of the equipment complement of the Rosboro mill:

> Power plant: three 600-horsepower boilers, a 1,000-KVA turbine generator, and a 500-KVA turbine generator, boosted by a capacitor to store power for high-demand situations.
>
> Sawmill: a 9-foot band saw with a 14-inch shotgun feed, a 10- by 60-inch edger, a 32-foot swing cut-off saw trimmer and a 32-foot automatic timber trimmer located behind the 10" edger, a 7-foot band resaw, a 48-inch pony

> edger, and a 20-foot automatic pony trimmer. Sawmill capacity: 125,000 board feet per eight-hour shift, depending on size of lumber.
>
> Dry kilns: six 14- by 114-foot kilns with automatic stackers and unstackers and a smooth-end trimmer. Dry-kiln capacity: 90,000 to 100,000 board feet per day.
>
> Planing mill: Two large planers with sorting chains, one of which was equipped with an overhead sorting rack.

The mill also had storage sheds for rough and finished lumber, a long loading platform, three lumber carriers for transportation, a 10-acre log pond, and a 60-foot wigwam refuse burner.[36]

The projected payroll for the first week of April 1940 came to $2,141, which cover the wages of eighty-four men. Two members of the seven-man electrical crew were the highest-paid, earning $1 an hour. For the others, wages ranged from 85 cents an hour for skilled blacksmiths and planer, sawmill, and dry-kiln operators, to 50 cents an hour for general workers. The managers and office staff collectively were to be paid $2,663 a month, or about $660 a week on average.[37]

Pond, power plant, sawmill, and waste burner

Rosboro's wages were probably lower than those of other local companies. Springfield Plywood, for instance, was offering its plant employees 65 cents an hour by the early 1940s.[38] Rosborough may have been a fair man, but he was not overgenerous. "As regards our wage scale," he wrote to Cole

Men of the Rosboro crew in the early 1940s

before the plant opened in 1939, "do not let anyone know anything about what wage scale is. If they do not have any more faith in my fairness than this, let them stay in Arkansas. There is a lot of men here that is just as good as we have there that will fit into the picture perfectly."[39]

Rosborough had operated in the non-unionized South for thirty years, and he was used to tractable employees. The Pacific Northwest, in contrast, had had an active labor movement since the 1880s, when immigrants had arrived from Europe with union experience and worked to organize mill and woods workers and others.[40]

In the early twentieth century, owners of the mines, mills, woods, and farms depended heavily on young, unmarried men for labor. Up until the 1930s, loggers and mill workers lived mostly in company-provided towns or camps, where the accommodations, food, pay, and working conditions could range from adequate to awful. In this volatile economic environment, young, footloose men made good workers, because they didn't have families to support. By the same token, however, the uncertainty and frustration of their lives made them ripe for union activism.

The union movement of the time flowed in two broad streams. Skilled workers could join the craft-based unions making up the American Federation of Labor. Unskilled workers, initially not welcome in the AF of L, found solidarity and a spirit of militant activism in industry-wide unions, the most

Log pond

famous of which was the Industrial Workers of the World.[41] The IWW, also known as the Wobblies (an epithet that its members adopted and wore with pride), was composed of the most alienated workers from the edges of society—miners, loggers, farmhands, factory workers, and mill workers. The IWW made its creed clear in its constitution: "The working class and the employing class have nothing in common."[42]

Through their early existence the IWW battled AF of L accusations that they were led by communists.[43] The Wobblies' tactics eschewed political action in favor of dramatic demonstrations on behalf of free speech and free assembly, some of which led to violent clashes with employers and the police.[44]

The Wobblies lost most of their steam during World War I, but their spirit rose again in the formation of the National Lumber Workers Union in 1929. The NLWU carried on some of the old Wobbly demonstration tactics, but they were unable to bargain effectively with mill owners.[45] In the meantime, Roosevelt's New Deal was making things easier for unions by, among other things, establishing collective bargaining as a right under the National Industrial Recovery Act. That law was soon struck down by the Supreme Court, but the right of labor to organize was ultimately preserved in other legislation.[46] The Rosboro plant opened in 1939 as a union shop and has remained one throughout its history.

Courtesy of Lane County Historical Museum

Aerial photo of the Rosboro mill, c. 1942

Up and Running

A flurry of equipment orders and bills in the winter and spring of 1940 testify to the quickening pace of construction on the Rosboro mill. After he had shipped and billed for at least $2,500 worth of equipment, and probably before he'd been paid, supplier J. L. Hollis brought up Rosboro's demand that they share the markup. "Naturally," he wrote to Cole, "we want to do what is right about this matter but on the bulk of the items we have sold you, we were allowed very little commission." He went on to detail the slim percentages the company had made on the various items, mentioning a discount he'd wangled for Rosboro on its purchase of band saws, and adding with faint reproach, "We sold about $80,000 worth of Simonds, Hoe and Atkins products last year and had no complaint of the prices charged." The letter concluded,

> After you see the small amount we are making on the transaction …we hope you can see your way clear to pay invoices as rendered. But as we wrote

> you, we will do whatever Mr. Rosboro [sic] thinks is fair and we leave the matter up to him.
>
> Please treat this letter confidential.[47]

No record can be found of whether Rosborough insisted on a discount or agreed to pay Hollis the full invoiced price, but it is clear that the company, squeezed for cash, was dragging out payment for some of its orders. "We regret the delay in settling our account," Cole wrote to J. L. Hollis in June, "but due to some delay we had with the R. F. C. in receiving funds we were just not in a position to take care of this as we should have done."[48]

The mill was operating by this time, cutting 100,000 board feet of 1-inch and 2-inch lumber per eight-hour shift. "We think we will increase this to 120-125 M' when the crew and machinery are thoroughly broken in....," said Cole, adding with a touch of pride: "100 M' is a lot of 1" and 2" to cut in eight hours."[49] The first shipment was expected to go out by the end of June, and the planer ought to be operating by that time, too, Cole said.

The company began shipping lumber in July, and by November Rosboro was making money. Gross receipts that month came to $5,800, and Cole expected this to increase by a lot very soon.[50] Rosborough was making plans to rearrange the lumber production flow through the resaw and install another remanufacturing band mill. The improvements would cost about $25,000, Cole estimated in a letter to Jess Hollis, but would save the company twice that much in a year.[51] If relations between Rosboro and Hollis & Co. had ever been slightly strained, they were back to normal now, as the companies exchanged holiday pleasantries. Hollis sent Rosboro a gift of company-branded pencils.[52]

An article about Rosborough's new venture had appeared in the *American Lumberman*, a lumber-industry trade journal. Rosborough sent a copy to his old friend Hollis, who replied, "[I] read with interest about you and the new plant at Eugene. ... 'America' needs such men as you and Mr. Cooper and Mr. Atkinson and Mr. Cole to carry on. You are the kind that keep American going."[53]

Troubled Times

Correspondence during these years mostly tends to business, but occasional hints emerge of the goings-on in Europe and Asia. A fast-talking demagogue had emerged from the ruins of Weimar Germany, and in 1936 his armies invaded the Rhineland, lost by Germany to France in World War I. The German people need *lebensraum,* Hitler declared. Two years later he pledged to leave the Soviet Union alone; nine days after that, confident that the Russians would do nothing to stop him, he invaded Poland, and Europe was at war.[54] Japan had invaded China in 1937, after claiming that the Chinese fired on Japanese troops at the Marco Polo bridge near Beijing. Within six months the relentless Japanese army toppled Shanghai and Nanjing.

The power plant

Allusions to these troubled times can be glimpsed in the letters of Rosborough, Cole, and their business associates, such as this one from a salesman at Hollis & Co.: "I think it is wise to book this material [saws and chain] ahead with the uncertain conditions existing. Our Rubber Belt Factory advised us that Great Britain has purchased a million dollars' worth of Fire Hose in this country and they had taken part of the order—which means that our deliveries will be slowed up considerably."[55]

When the United States finally entered the war two years later, the Pacific Northwest—indeed, the whole country—experienced profound social and economic dislocation.[56] Men went off to fight; women went to the factories to do the men's jobs and then came home to do their own. Commercial shipyards were turned into vast war-production centers, and cities such as Portland and Seattle swelled with migrants from other parts of the country. Families lived with high prices, cramped quarters, rationing, and fear of losing loved ones. At the same time, the Second World War brought unity and solidarity to America, whose citizens mostly felt certain that they were sending their sons to fight and die for a just cause.

The war accelerated demand for steel, aircraft, lumber, food, clothing, and armaments. Even before the United States officially joined the fight, American goods were being shipped overseas to help friendly nations do battle with their, and our, enemies.[57] The United States was supporting the Chinese in their struggle with the invading Japanese by sending supplies along the overland Burma Road[58]—and then, when the Japanese closed the road in 1942, "flying the Hump" to China in C-47 transport aircraft.[59]

After the Japanese bombing of Pearl Harbor, demand for goods for the war effort skyrocketed, especially lumber, and that, following the bleak prewar years, was good news for Rosboro Lumber Company and the other mills of the Pacific Northwest.

Loadng a rail car

The new company was still struggling during the months leading up to Pearl Harbor. A March 1941 memo from Cole to Rosborough gave the news that "the Company had an abnormally high productive cost in the manufacture of its inventory" in the first half of 1941, mostly because they'd had to sell the products cheaper than they'd planned to cover cash flow shortages stemming from purchases of capital items such as mill equipment.[60]

By January of 1942, however, Cole cautiously characterized business during the past year as "fairly good" and said it might have been better had the mill not been under continued construction.[61] The mill, he wrote to Jess Hollis, now had a headrig, a pony rig to cut large cants, and a second, smaller resaw "to remanufacture anything that gets out of the saw mill that should be edged, ripped or resawed." The green chain had been upgraded with the addition of jump rolls. Production was now 140,000 board feet per 8-hour shift, and Cole was confident it would increase to one hundred fifty thousand once the crew was trained on the new machinery.[62] Hollis's reply, on March 17, 1942, mentions that the equipment-supply business was also good last year—about 30 percent higher than any previous year, "and quite a bit of the sales were to defense plants"—in fact, he said, about 40 percent of his business was going to the war effort and it was hard to get hold of merchandise to sell.

There was much demand for products made from the West Coast's clear old-growth timber. It was used for paneling and ship decking, and for hulls for minesweepers, among other products.[63] An article in the February 1944 *Timberman*, a lumber-industry trade journal, said Rosboro Lumber, "with over 40 million feet of production, last year has been one of the big contributors to the war program."[64]

Lumber operators felt themselves hampered by wartime regulatory mechanisms imposed by the federal government.[65] They particularly complained about wage limitations, which the government had imposed to control inflation. In addition, with most of the working-age men off to war, the mills were constantly scrambling for labor, and they made up the gap with older men, boys, and women.[66] Wartime wages in West Coast mills had shot up to an average $1.04 an hour by 1943.[67]

Rosboro Lumber was making a wide variety of products. It was in what was known as the "mixed-car trade"—that is, the mill was set up to manufacture many different products of various lengths, widths, and degrees of finish work. A railroad car might ship out with framing lumber, shiplap (a lumber sheathing item), paneling, siding, mouldings, vertical and flat grain clear items such as flooring, ceiling patterns, and other products. "You could literally build a house with what was in one shipment," said Louise McQuin.

Such a variety of products required a lot of manpower in the mill. Like its counterparts in the industry, Rosboro coped with the labor shortage by hiring older men and boys and augmenting its crews with seasonal help during Christmas and summer holidays. Unlike some of the large war industries like the Kaiser shipyards in Portland, and even some of its counterparts in the lumber industry, Rosboro did not break the gender line on its factory floor. Henry Kaiser may have had his Rosie the Riveter, but there were no Gerties on the green chain at Rosboro. To Paul Cole's recollection, no women worked in the plant until about 1970.

Cole, by this time a student at University High School in Eugene, was one of the company's seasonal employees. He was a good student, but not much interested in the lumber business. He had a mechanic's head and an artist's spirit, and he thought he might like to be an architect. Believing his son needed some discipline, the elder Cole arranged for him to work in the company office during his summers off.

His first task was learning lumber invoicing from Louise McQuin, who was an expert. Invoicing was a complex job because the sawmill made so many items. She taught Cole to be careful. "We checked each others' work on these invoices," he said, "because if there was a mistake in the customer's favor they would not tell us." He also took the bills of lading to the Southern Pacific office each day.

Other summers he worked in the planing mill—he was underage but legal, because the war was on. Sometimes he helped sales manager Al Carley, working on inventories, writing letters to customers, selling lumber on the telephone, making quotations, and trying to find new customers.

One summer Paul worked in the woods operation with Bigley and Feiss Foresters, who were mapping and cruising the company fee timber. The cruising crew was composed of Oregon State College students who were trained in this kind of work. This put Paul at the bottom of the heap, in the go-fer position. "We located section lines, did topography work, chain work on lines, etc.," he said. "My job was always assisting them, since I had no forestry-type education."

Back Home

As the war dragged on, Whit Rosborough was starting to feel his age. He'd done well for himself in this late-life venture. He'd paid off all his debts; indeed, he'd become a millionaire. But he was worried what would happen after the war was over. What if there was another depression like the one that had stricken the country after World War I? He was also starting to have heart trouble.

But mostly, he was homesick. There comes a time in life, he later mused, "when a man's heart is with the friends of his youth and his kinfolk."[68] He was going back home.

Notes

1. Letter from T. W. Rosborough to B. S. Cole, May 25, 1939.
2. Interview with Claude Gerlach, August 9, 2005.
3. Velasco, Dorothy, and Mara Velasco. *Springfield: Between Two Rivers* (Montgomery, AL: Community Communications, Inc., 1999), p. 64.
4. Display at Springfield Historical Museum, Springfield, OR, viewed July 7, 2005.
5. Ibid.
6. City of Springfield Mill Race history web site, http://www.ci.springfield.or.us/Pubworks/MillRace/history.htm, consulted Dec. 6, 2006.
7. The power plant eventually became a part of Pacific Power and Light Company, which vacated the building in 1976. It escaped demolition and became the home of the Springfield Museum in 1979. Ibid., Velasco and Velasco, p. 145.
8. Schwantes, Carlos A. *The Pacific Northwest: An Interpretive History* (Lincoln and London: University of Nebraska Press, 1989), p. 302.
9. Ibid., Velasco and Velasco.
10. Ibid., Velasco and Velasco. Rosborough was aware of the excellent fishing opportunities: "Eugene, Oregon has an excellent summer climate and I am told that the fishing is good." (Letter from T. W. Rosborough to Milo T. Jones, Feb. 14, 1939.)
11. Ibid., Velasco and Velasco. The description is somewhat lackluster, but what can you say about a movie theatre?
12. Springfield *News, Special Enterprise Edition*, about 1939.
13. Letter from B. S. Cole to Nelle Beidleman, July 13, 1939.
14. Letter from B. S. Cole to J. L. Hollis, Dec. 11, 1939.
15. Springfield *News, Industrial Progress Edition* (undated; some time in 1939).
16. Ibid., Springfield *News, Industrial Progress Edition*. The article said "The Springfield industrial tract was available without cost for construction of the industry, but Mr. Rosborough preferred to buy the site outright from the city."
17. Letter from J. L. Hollis to B. S. Cole, Aug. 4, 1939.
18. Letter from Joe D. Royer to B.S. Cole, Sept. 25, 1939.
19. Memo by B. S. Cole, Sept. 20, 1939.
20. Letter from B. S. Cole to J. L. Hollis, Dec. 11, 1939.

21. Ibid., Cole to Hollis, Dec. 11, 1939.
22. Ibid., Cole to Hollis.
23. Letter from J. L. Hollis to B. S. Cole, Dec. 16, 1939.
24. Letter from B. S. Cole to Chas. Hilburn, Oct. 10, 1939.
25. Letter from B. S. Cole to Sam L. Smith, Feb. 13, 1940.
26. Letter from C. J. Hilburn to Rosboro Lumber Co., Oct 10, 1939.
27. Interview with Clyde Hughes, Oct. 14, 2005.
28. Letter from B. S. Cole to J. D. Royer, Oct. 4, 1939.
29. Letter from B. S. Cole to Sam L. Smith, Jan. 26, 1940.
30. Ibid., Springfield *News Industrial Progress Edition.*
31. Letter from Sam L. Smith to B. S. Cole, Jan. 26, 1940. The creative spelling is his.
32. Ibid., Cole to Smith, Jan. 26, 1940. It is unclear why these letters have the same date. This letter seems to be the query, to which Smith replied in a handwritten note. Maybe Cole's letter was hand-delivered? Or maybe Smith was at the plant and was replying to questions put to him orally? Or maybe there was a phone conversation before Cole's letter, or maybe Cole dated the letter a day ahead.
33. Report by Herbert Hessler, operating consultant, April 13, 1940.
34. Ibid., Hessler.
35. Ibid., Cole to Hollis, Dec. 11, 1939. One can almost hear the sigh in his voice.
36. Ibid., Hessler.
37. Ibid., Hessler.
38. Ibid., Velasco.
39. Ibid., Rosborough to Cole, May 25, 1939.
40. Thoele, Michael. *Bohemia: The Lives and Times of an Oregon Timber Venture* (Portland: Oregon Historical Society Press, 1998, p. 144).
41. Ibid., Schwantes, p. 261.
42. Ibid., Schwantes, p. 261.
43. Ibid., Thoele, p. 144.
44. Ibid., Schwantes, p. 262. Schwantes adds (p. 263) that the Wobblies were more often victims than perpetrators of labor-related violence, which had erupted periodically during the two decades before they had appeared on the scene. "Rather than initiate a new and violent era of labor-management relations," Schwantes writes, "the IWW merely elaborated on a tradition of militance and radicalism that already existed among Northwest wageworkers."
45. Ibid., Thoele, p. 144.
46. Ibid., Thoele, p. 145.
47. Letter from J. L. Hollis to B. S. Cole, April 27, 1940.
48. Letter from B. S. Cole to J. L. Hollis, June 22, 1940.

49. Ibid., Cole to Hollis, June 22, 1940.
50. Letter from B. S. Cole to J. L. Hollis, Dec. 24, 1940.
51. Ibid., Cole to Hollis, Dec. 24, 1940.
52. Ibid., Cole to Hollis, Dec. 24, 1940.
53. Letter from J. L. Hollis to T. W. Rosborough, Dec. 3, 1940.
54. From Ad Access web site, http://scriptorium.lib.duke.edu/adaccess/timeline-1936.html, consulted Oct. 24, 2005; also from About web site, http://history1900s.about.com/library/weekly/aa022500a.htm, consulted Oct. 24, 2005.
55. Letter from Joe B. Royer to B. S. Cole, Sept, 25, 1939.
56. Ibid., Schwantes, pp. 326-8.
57. Ibid., Schwantes, p. 328.
58. From web site The Japan-China war, 1937-1945, http://www.historylearningsite.co.uk/china_war.htm, consulted Oct. 25, 2005.
59. *Journal of the Air Force Association* web site, http://www.afa.org/magazine/1991/0391hump.asp, consulted Dec. 6, 2006.
60. Memo from B. S. Cole to T. W. Rosborough, March 12, 1941.
61. Letter from B. S. Cole to J. L. Hollis, Jan. 21, 1942.
62. Ibid., Cole to Hollis, Jan. 21, 1942.
63. Interview with Paul Cole, July 7, 2005.
64. "'You're a Goner if the Sawmill Bug Bites You,' says T. W. Rosborough" (reprint from *The Timberman* 15(4), February 1944.)
65. Ibid., Thoele, p. 190.
66. Ibid., Thoele, p. 190.
67. Smith, Kenneth L. *Sawmill: The Story of Cutting the Last Great Virgin Forest East of the Rockies* (Fayetteville, AR: University of Arkansas Press, 1986), p. 196-7.
68. Eugene *Register-Guard*. "*T. W. Rosborough, of Texarkana, Springfield, Eugene*" (editorial May 7, 1952), cited in Smith, *Sawmill*, p. 188.

3

ROSBOROUGH GIVES OVER THE REINS

"My father had the idea that the yield could and should be sustained. I, too, have that idea."

—Paul Cole

AFTER A SHAKY start, the new Rosboro Lumber Company prospered through the early years of the Second World War. In fact, the war had a lot to do with that prosperity, for Rosboro and the whole West Coast timber industry. The war brought a stream of lucrative orders—plywood for construction of barracks and other war-related structures; for ship decking and minesweeper hulls (because wooden-hulled ships would not set off the magnetically triggered underwater mines), for lumber to expand military bases, spruce for airplane wings, and for all manner of other wartime supplies. The war brought secure employment for loggers and mill workers,[1] whose ranks were thinned when many men went to war or to work in shipyards and other war-related factories. Said Ruth Cole Kainen, Paul Cole's sister: "Mr. Rosborough would have gone broke if it hadn't been for the war."

But Whit Rosborough, remembering the bad years after World War I, didn't think the prosperity would last, and besides, he was getting restless, ready to move back to Arkansas and retire. He was an entrepreneur, not a manager—more interested in building a company than running one.

Quietly, he sought out potential buyers for his company. One of his prospects was Weyerhaeuser Co., which had owned timber in Lane County since 1907[2] and had a plywood plant in Springfield.[3] That inquiry evidently came to nothing.

Ultimately Rosborough proposed to sell the company to three of the men who'd been with him from the start of his Oregon venture: Beuford Cole, his general manager; Tom Watts, the superintendent of the mill; and Vernon Williams, the logging manager. Joining them was a fourth partner: Spencer Collins, a Eugene accountant who was handling Rosboro's taxes and other financial matters. Along with his expertise with numbers, Collins brought a $250,000 investment into the partnership—a pretty big sum in those days. Whit Rosborough, however, was the chief financier of the partnership.

Thomas Whitaker Rosborough

Cole, Rosboro's treasurer, had been given the title of general manager in 1943. He had been acting in that capacity since the very first days of the company. Cole was not in a position to finance anything, but his expertise in mill construction and equipping, timberland purchasing, personnel management, and government relations, along with his deft handling of the myriad crucial details of building a company, would prove invaluable to Rosboro's ultimate success.

Cole may have doubted his own value from time to time, according to his daughter Ruth. "He always had the idea that he could never get another job—that he was tied to Mr. Rosborough." Cole's wife, Teden, had made it known that she was unhappy with her husband's long hours and low pay. Ruth Kainen recalls some tense moments between her parents, although, "being sixteen, I was… not all that conscious of the strain in the family." She knew her father had a hot temper; the reason for some of his ire, she said, is clearer in hindsight.

One night during the time that the details of the transfer were being discussed among the company's principals, Ruth recalls hearing her parents quarrelling loudly. Soon after that, Beuford Cole went to see Spencer Collins, asked for a job, and was offered one, at a higher salary than Rosborough was paying him. "During this time my father was also offered other general manager–type jobs by other lumber firms in the area," said Paul Cole. "Rosborough heard of these jobs and knew he had to keep [Beuford Cole] or he was going to be in big trouble."

"Father went in and told Mr. Rosborough that he had to quit because his wife was unhappy," said Ruth Kainen. "Mr. Rosborough made him a counteroffer and made him a partner." She speculates, and her brother Paul agrees, that Beuford Cole would not have been offered a partnership in the new company "if Mother hadn't goaded him into it."

In his memory of that time, Paul Cole is more sympathetic to Rosborough's position. "I know Mother had a low opinion of Whit Rosborough because she thought he didn't pay my father enough. Well, part of running a business is you pay what the market will bear. I can understand the reasoning Rosborough used for not paying Dad more. As an employer, I might have done the same in his circumstances."

Nevertheless, in the end Cole became one of the four partners who assumed ownership of Rosboro Lumber Company in May of 1945. To effect the transfer of ownership from Whit Rosborough to the partners, Rosboro's board of directors created a shell company, Rosboro Timber Company, which then sold the assets, including timberlands and facilities, to the new partners, who were operating under the old name of Rosboro Lumber Company.

The players in these two corporations were the same people, of course, including Whit Rosborough, who had owned most of Rosboro from the start. The corporate maneuvering allowed Rosborough to essentially finance the sale of the company himself, although there was other financing as well: the money from Spencer Collins, as well as contributions from two large customers, the Hardman Lumber Company of Kansas and the Peter Kuntz Lumber Company of Ohio, and some bank loans.[4]

The new office on 25th and Main

Under the terms of the transfer, the new partners would lease the sawmill and all attendant equipment for fifteen years at $180,000 a year. They could buy whatever timber they needed from the Rosboro holdings at fair market value. In addition, they would assume the other assets of the company, including the log and lumber inventories, trucks and other equipment, and accounts receivable, at a value to be determined by accountant Spencer Collins. They would pay this sum off within a year at 4 percent interest.[5] The accounts receivable included $28,000 owing from a company called Clark and Logsden. The partners also assumed the accounts payable and a couple of other debts, including a note for an unspecified amount held by Mattie Holliday, one of the stockholders,[6] and some money owing to March Logging Company.[7]

In addition to the yearly lease payment, the immediate bill to the partners came to $270,070.06. They were not required to put up collateral, because "the partnership is financially responsible for the payment of this note and the performance of all contracts and obligations made by it."[8]

Rosborough had been essentially retired since 1944.[9] In December of

1946 he resigned as president and director of Rosboro Timber Company, and his position was not refilled. After the successful transfer of the company to its new owners, Rosboro Timber Company was dissolved.[10] Whit and Anna moved back to Texarkana, where they built a big house with an adjoining apartment for his long-time household staff members, Fred and Halcy Wingfield.

Whit Rosborough enjoying a relaxed moment

Although he was nearly eighty, Rosborough's empire-building urge had not left him completely. Back home, he organized the Ark-La-Tex Lumber Company and started building a mill near Texarkana.[11] "When TWR got into trouble with the new sawmill," recalled Paul Cole, "my father went to some effort to help him in every way possible."

During this time, according to one reminiscence, Rosborough remarked to Beuford Cole, "I may lose my money, but I'll have fun!" Cole replied, "You won't have much fun if you lose your money."[12] In fact, he lost his health first, and was forced to sell the mill at a large loss before it had even opened. He died in 1952 at the age of eighty-three.[13]

Ruth Cole Kainen offers a succinct obituary for Thomas Whitaker Rosborough: "He lost a million dollars, and then he died. I believe the two things were connected." Margaret Rieder's recollections of her uncle, however, indicate that he lived a rich, full life, with many accomplishments and few regrets. "He was fun. He had a great sense of humor. People enjoyed being around him. He really enjoyed life, he and Aunt Anna both."

Scrimping for Success

In the summer of 1945 the partners owed $200,000 on the $270,000 note for Rosboro Lumber Company's assets,[14] plus $180,000 for the yearly lease. They were in deep debt—"in hock up to *here*," said Hub Lindsey, Martha Watts's nephew. The first few years were a financial struggle. The partners scrimped to make the payments and took far less in salary than they might have wished.

Tom Watts, dwarfed by huge logs

They were in some ways odd bedfellows. Tom Watts was in because of his family connection to Rosborough: his wife, Martha, was Rosborough's niece, the daughter of his sister Nelle Beidleman; and his sister, Anna Watts Rosborough, was Rosborough's wife. Tom knew a lot about sales—he'd been in tires back in Arkansas—but not much about the lumber business.[15] Rosborough had made him superintendent of the sawmill, a job for which he had little experience.[16] "Whit Rosborough had a habit of putting people into jobs that they had no preparation for," said Paul Cole. "From my viewpoint, Tom was okay in the job because he had capable people running the various parts of the sawmill operation. Tom and Martha were very much a key to the transition from TWR to the succeeding partnership."

Martha Watts, Anna and Whit Rosborough, and Tom Watts

The Wattses evidently had some means, for they had lived in relatively high style in Arkansas. Tom drove a nice car, and Martha dressed stylishly, her penchant for showy jewelry evident even in those early years. "Tom and Martha would come to Huntsville to see my daddy," recalled their nephew Jim Watts, Jr., the son of Tom's brother, James Foster Watts. Young Jim Jr. was still in grade school at the time. "They always had the nicest of vehicles, and I remember once that Martha had a suitcase, or purse, I guess it was, made of clear plastic. You could see right through and see the contents of it."

This impressed the young man mightily. "She wore very nice, large jewelry," Watts added, "and it was *not* rhinestones, if you know what I mean." Childless, Tom and Martha Watts were generous with their nieces and nephews on both sides of the family.

By several accounts, Martha had a sharp head for business. "She had outstanding intuition," said Jim Watts, "as well as knowledge and experience." Martha Watts would prove to be a valuable resource for Paul Cole when he assumed the reins of the company in 1958. "She was easy to discuss things with," he said, "and quick to understand the many issues I talked to her about."

Vernon Williams, who left the partnership early, seems to have left a lighter impression on company records and on people's memories. His father, L. D. Williams, had hired on with Rosborough's operation in Arkansas in about 1907 to oversee construction of a logging railroad, and had stayed to become logging superintendent for the Caddo River Lumber Company for the rest of his career.[17] Son Vernon, trained as a civil engineer and land surveyor,[18] had been working for Whit Rosborough as logging superintendent at Forester, Arkansas,[19] when he accompanied his boss out to Oregon in 1939. He managed the timberlands for the new Rosboro Lumber Company. His biggest job, and quite an achievement by Paul Cole's reckoning, was to build an entire logging-road system on the Finn Rock tree farm.

Roy Edwards, Rosboro's long-time head sawyer, remembers Williams as "hot-tempered" and "hard to get along with." Edwards was Williams's brother-in-law—the two wives were sisters. In fact, it was at his sister's invitation that Edwards and his wife had come west from Arkansas in 1940 to work for the new sawmill in Springfield.

Vernon Williams stayed with Rosboro until 1951, when he and his wife moved back to Arkansas. He sold his share of the company, amounting to 7 percent, to partners Anna Rosborough and Spencer Collins, Martha Watts's sister Mary B. Lindsey, Mary's son Hubert (Hub), and Beuford and Teden's son Paul. At the same time, the elder Coles made things equitable among their children by selling the same percentage of their interest to their daughters Elizabeth and Ruth. Thus part of Rosboro Lumber Company passed for the first time into the hands of a younger generation, setting the pattern for Rosboro's remaining a family-controlled company throughout its history.

Spencer Collins is remembered as something of an odd duck, a truly eccentric person—"unusual" and "different," in the words of the understated Paul Cole. Cole's wife Nonnie recalls him as "an enigmatic man." The only Oregonian in the group, Collins was a partner in the first, or one of the first, accounting firms in Eugene. He lived in an apartment above his office on 11th and Pearl Streets. "He was a big, tall fellow, white hair, rolled his own cigarettes and smoked them in a holder. When he'd get to talking, he'd spill his tobacco," recalled Paul Cole.

Collins's accounting and financial expertise were essential to the partnership's success, and his quarter-million-dollar investment didn't hurt. Within a few years he and about ten others, including Beuford Cole and Tom Watts, would found Citizens Bank of Oregon, on which the company relied for some of its later financing needs—for example, approximately $2 million for the veneer-plywood plant the company built in 1959-60. Beuford Cole was chairman of the board of Citizens Bank, and Paul Cole became an active director after his death in 1958.

Logging operation at Finn Rock tree farm

Beuford Cole's contribution to the partnership was his utter indispensability: the other partners could not have run the company without him. Socially, the Cole family was, or felt itself to be, in a different stratum than the Rosboroughs and the Wattses. Beuford Cole had been raised poor in Mississippi and had gone to work at age twelve to support his family. Teden Cole had a high-school education, quite respectable for a woman of her day. Her husband had less formal schooling than she, and he drove himself diligently throughout his life to educate himself. According to their daughter Ruth, Beuford and Teden felt snubbed by their business partners and by the

Eugene society in which they had hoped to find a place. It didn't help that Oregon was chillier, in all senses of the word, than the warm South. "This was a great disappointment to my father," said Kainen. "I felt he blamed my mother, but I'm not sure what she could have done about it."

Paul Cole's take on the situation is that his mother was perhaps too sensitive. "I believe this was all about my mother's feelings. My father was a very active man in his business and social life, was on various charitable and industry boards of directors, and had a lot of business and personal friends." He acknowledged that his parents "were completely different from Martha and Tom Watts in the social arena," with very different cultural backgrounds and modes of upbringing. "Mother was a daughter of a poor farmer in Arkansas, whereas the Wattses were from cities, not poor, and had a lot more experiences in all phases of life."

Teden Cole "was an unhappy person," said Paul's wife Nonnie Cole, who acknowledges that she and her mother-in-law did not get along well. "She was the kind of person who is ill at ease in any situation, and who makes everybody else ill at ease."

Surprising Prosperity

When the war ended and wage and price controls were lifted, Rosboro and other West Coast mills began to do very well. The postwar prosperity surprised many—T. W. Rosborough could be forgiven for his pessimism. "Much of the planning for the post-war period," said a 1948 report from the Oregon Department of Forestry, "was based on the possibility of a return to unemployment, made-work and large relief expenditures by government agencies." However, "Three years of post-war readjustment is now history. Instead of unemployment and low commodity prices, there is the greatest per cent of the population working ever known and also the highest commodity prices on record."[20]

Here is how another Oregon lumber company saw the situation:

> ...the huge, pent-up, war-restricted demand for lumber surged over the industry in an uncontrollable flood. As a consequence, prices late in 1946 became chaotic; ... Builders with partially completed homes needing upper grade lumber for finishing, stormed the mills with fantastic offers of price and specifications to cover their needs. Industrial lumber users denied raw materials for making their products by price or priorities, clamored for their share of

> the available supply. Lumber hungry foreign nations importuned their American agents to get some lumber to them at any price.[21]

Given the favorable climate, it is not surprising that the Rosboro partnership turned in a respectable performance in its first half-decade. By 1950, Rosboro Lumber Company was producing about 40 million board feet of lumber a year,[22] sawing logs from the big old-growth Douglas firs, hemlocks, and cedars on its township-sized tract up Quartz Creek, in the McKenzie drainage. These logs yielded 30 to 40 percent clear lumber–i.e., without knots or defects.

Felling a big tree (without benefit of hard hat)

The company's product line continued to be diverse: dimension lumber, both green and kiln-dried, as well as timbers, cants, house siding, ship decking, heavy flooring and roofing, molding, window and door stock, and stock for furniture, airplanes, and caskets. After 1951 Rosboro also produced wood chips for kraft and other paper products. Sawdust and bark were used to fire the three boilers that produced steam to run the sawmill's two steam shotgun carriages, as well as the dry kilns.[23]

In 1950, after graduating from Willamette University with a degree in business, Paul joined the company full-time. He started in the sales office, working for sales manager Al Carley in the same position as he'd had the previous summer.

Now, however, things were different. Paul was soon promoted to assistant general manager, and it was understood that he was being groomed for the general manager's position after his father retired. Roy Edwards, by then working as pony-rig sawyer, remembers Cole as a young man with a bright future. "He'd worked in the ...mill for a while," he said, "but he wasn't a mill worker. Running a company—that was Paul's nature in life. He's smart—very smart."

In December of 1950 Paul Cole married a Willamette classmate from Honolulu, a pretty girl with a winsome smile. Her name was Norma Lucia, but everyone called her Nonnie. The following year, at his father's urging (and with his financial help), Paul Cole bought some of Vernon Williams's interest in the company as Williams prepared to move back to Arkansas. Paul and Nonnie bought a big lot near Hendricks Park in Eugene, looking east over the river to Springfield, and built a new house. In 1952 their daughter Debbie was born; she was followed by Phoebe in 1954.

The company continued to prosper. "When I used to come out and visit in the '40s," said Hub Lindsey, "Uncle Tom and Aunt Martha were driving a Chevrolet." In the '50s they were driving a Cadillac. The Wattses built a fine new house on the site of their old one in Eugene, near the University of Oregon. They liked the neighborhood, so they had the old house moved away and built the new one on the same site. Martha Watts "was quite the doyenne of society," recalls Nonnie Cole. "She had wonderful clothes, very tailored. A handsome woman."

Constructive Competition

Beuford Cole, who had remained general manager after the partners took over, was fifty-eight years old by then, as vigorous and work-focused as ever. Through the late '40s and early '50s he continued to buy timberland, adding to the company's already extensive holdings. He also found buyers for Rosboro logs that were less than suitable for its mill, and he bought logs from other companies and, increasingly, the Forest Service. He had to watch markets carefully in negotiating such deals, for sometimes it was more profitable for the company to keep its own logs and saw them into lumber, but sometimes the raw logs brought the highest value. Sometimes government timber would be worth bidding on but sometimes it wouldn't.

It was under Beuford Cole that the web of productive relationships among Rosboro, the Forest Service, and the company's competitors was forged. Such relationships add up to a curious symbiosis among the players in the deeply connected Northwest wood-products industry. Timber companies are competitive, but they are cooperative, too, for they are close neighbors, both in the woods and in the communities where their mills are located. Company A might own a tract of forest that is closer to Company B's logging road; Company B might have logs that are more suitable for Company A's mill. Many deals are cut, and the trading is often sharp.

An example of this is found in the sustained and complicated negotiations between Cole and the Weyerhaeuser Co. between 1941 and 1952, involving land, logs, and eventually the sale of a Rosboro road to the Forest Service.

Weyerhaeuser owned a section of the McKenzie River drainage, Section 16,[24] which was bisected diagonally by Rosboro's logging road. Beuford Cole had negotiated the easement for use of the road in July of 1939 with the Drew Timber Company. Weyerhaeuser aquired the easement when it bought the Drew timberlands shortly thereafter.[25]

In 1941 Beuford Cole started negotiating with Weyerhaeuser's timber manager, Minot Davis, to buy and log the timber in Section 16 and part of the adjoining Section 15, which Weyerhaeuser also owned. Weyerhaeuser was agreeable to this, said Davis, provided Rosboro delivered all the fir "peeler" logs to Weyerhaeuser at the Springfield Plywood Corporation, a Weyerhaeuser subsidiary. The timber would cost Rosboro $3 per thousand board feet.[26] If Rosboro didn't want to sell back the peelers, he said, Weyerhaeuser wouldn't sell them the timber.

Weyerhaeuser would also like to buy peeler logs from Rosboro's land, said Davis. Any chance of that? Cole replied that Rosboro would prefer to keep the peelers from its own lands. Although the company did not manufacture plywood or veneer, these fine logs also made high-quality flat and vertical-grain clear and upper-grade lumber. Rosboro would certainly be willing to log Weyerhaeuser's timber and sell the peelers to Springfield Plywood, although "we consider $3.00 a little high, but you own the timber and we suppose this is your price at present."

After some haggling over the down payment, the deal was struck. In his letter accompanying the signed contract, which called for Rosboro to be responsible for the taxes on the logged land, Beuford Cole suggested that it might be decent for Weyerhaeuser to give Rosboro title to the land in Section 16 after it was logged—just give them the land, since it was no longer useful to Weyerhaeuser. "We have bought some other timber that was handled in this way," wrote Cole."[27]

Bucking a big log

Davis replied that the company would consider deeding over the land once it was logged, but only if Rosboro would buy another portion of the company's holdings, Section 36, which had not yet been logged.[28] In his reply, Cole backed away from his request: Rosboro wasn't in a position to buy Section 36 and therefore wouldn't insist on getting title to Section 16; he was only floating the idea because he "rather thought this was customary."[29]

Rosboro finished logging Section 16 in late 1943. Cole wrote to Davis and offered to buy Section 36, but by then Weyerhaeuser was unwilling to sell. Good timberland had become quite a bit more valuable in the intervening two years.[30] "Ordinarily," wrote Davis, "we are more interested either in making exchanges or in holding our timber until such time as it is actually needed by some operator."[31] He didn't suppose Rosboro would be interested in a land trade? No, Cole replied; Rosboro didn't have any land suitably located, and the company was not in a position to buy the timber alone (separate from the land) from Section 36 right now.[32]

In 1948, Cole again picked up the thread of the conversation with Weyerhaeuser. Now there was a new complication: the Willamette National Forest wanted access to Rosboro's road across Weyerhaeuser's Section 16. The Rosboro road was the only truck access to the Forest Service land that lay behind Rosboro's holdings. The Forest Service couldn't sell any of that timber without a guarantee that the purchaser could get to it.

Rosboro owned the road but not the land. It had only an easement from Weyerhaeuser, and a second-party easement was not enough to suit the Forest Service.

Cole tried once again to get Weyerhaeuser to sell Section 16. This time he was dealing with local mill manager Jon Titcomb and Charles Weyerhaeuser. Weyerhaeuser offered to sell not the land but a permanent right of way, which he judged would suit the Forest Service better than a temporary easement.[33] But the Forest Service, as it turned out, wanted even more: it wanted title to the road, with the eventual goal of turning it into a public Forest Service road.[34]

This was a disturbing development. Either Rosboro had to buy the land from Weyerhaeuser at whatever price Weyerhauser might set, or it had to surrender the road easement and let the Forest Service negotiate the purchase with Weyerhaeuser. Either way Rosboro would not enjoy much of a bargaining position.

"It would seem, from our conversation with them [the Forest Service], that several lumber companies have executed a deed of this sort [to sell a logging road] recently," he wrote to Charles Weyerhaeuser. However, he pointed out, these were companies that had already cut most of their timber and were ready to move out. Rosboro, in contrast, had plenty of timber left in the area, and would really rather not sell its road to the government.

Cole also reminded Charles Weyerhaeuser of Rosboro's recent contract to log Section 36, and said the company planned to start logging that fall,[35] again delivering the peeler logs to Springfield Plywood. The price had gone up, however, because plywood peelers were in high demand. Cole pointed out that Springfield Plywood was paying Booth-Kelly not $3 per thousand board feet, but $3 per thousand *above* the going market price. He suggested that the Rosboro-Weyerhaeuser contract be revised to reflect the same price premium. If Weyerhaeuser objected and subsequently decided it didn't want to buy peelers from Rosboro, that was fine with Cole, "as we like to cut high grade logs in our mill."[36]

The question of buying Weyerhaeuser land in Sections 16 and 36 remained open as Rosboro negotiated the road easement with the Forest Service. Cole took up the matter again with Weyerhaeuser in 1951. Back in 1941, he reminded them, they'd seemed willing to deed the cut-over land in Section 16 to Rosboro for free or for a nominal sum. To be sure, conditions had changed, he acknowledged, but perhaps Jon Titcomb would be so kind as to "take the matter up with his people and see if he could not get them to agree to sell the land to us."[37]

The Rosboro archives have no record of Weyerhaeuser's offer, but Cole's reaction to it was outraged: "The price you made to us is so far in excess of anything we had ever expected to pay that we think we will have to reexamine and reconsider the entire situation..." Weyerhaeuser's asking price, he said, was unfairly based on the expectation of future tolls from the government timber that would one day flow along the road. "We had expected to pay you full value for the land itself and any relogged timber on the land, but not more than [that]."[38]

Nevertheless, Rosboro eventually made the purchase, and in 1952, after prolonged negotiations with officials at the Willamette National Forest, the company agreed to a permanent twenty-five-year easement for use of the road. The easement provided that in 1975 the title would revert to the Forest Service and the road would become a public forest road.

It would be interesting to know whether Cole succeeded in negotiating a price from Weyerhaeuser that he thought was fair, and whether the transaction warmed or cooled the relationship between the companies. Unfortunately, the archives offer no clue, but in Paul Cole's assessment the relationship went on pretty much as before. "Business relationships don't necessarily change because of differences in opinion or financial decisions," he said. "The general test is this: if a deal is a good one for the companies involved, they will still talk. Despite the posturing, in the final analysis, it's just business."

Into the Modern Age

In the years after the Second World War, America was growing an industrial and commercial enterprise the likes of which had never been seen before. Companies were expanding nationwide, and their management was becoming more centralized and hierarchical. In *modus operandi* and spirit, American business was modeling itself along the lines of the military enterprise that had just won the biggest war of all history.

"This was a long time coming," Paul Cole said, "due to owner-manager type businesses where the head person feels he has to make all of the major decisions. I believe Rosboro was far ahead of our competitors in this regard. We used a chain-of-command system, but with managers invested with both the responsibility and authority to run their segments of the business." At Rosboro this arrangement was an essential step in the evolution of the team-style management that exists today.

The forest-products industry as a whole was in the midst of a transition. As the old-growth timber was disappearing, the old mentality of "cut out and get out" was dying. If the industry wanted to survive in the modern world, its leaders knew they needed to conserve and sustain the forest resource into the indefinite future. That meant long-term planning and increased efficiency in the mills and the woods. No longer could operators afford to leave cut-over or burned-over lands to the uncertain efficiency of nature, or abandon them to counties for non-payment of taxes. Reforestation had to become a top priority.

In Oregon, research to support these new ways of doing business was coming out of the Oregon Department of Forestry, the School of Forestry at Oregon State University, and the Forest Service. The Forest Products Laboratory opened at Oregon State in 1947.[39] Its research—supported partly by a tax on timber harvests—included studies on reforesting of cut and burned timberland through both direct seeding and planting of seedlings. This work was prompted in part by the massive Tillamook fires of 1933, 1939, and 1945, which burned more than 300,000 acres of mostly virgin forests in northwestern Oregon. Other studies focused on efficient methods of "relogging"—going back into a site where the old-growth trees had been logged and cutting the younger trees before reforesting the site.[40] There was also much research on utilization of waste wood, especially bark, and on protection of timber from fire.

State law and the tax system also were evolving from the old "cut out and get out" days. In 1941, Oregon passed the nation's first law requiring that operators provide for reforestation of cut-over lands. The law was revised periodically to tighten the reforestation requirements and to give the state greater enforcement authority.[41] In addition, the method of taxing forest lands was changed to defer most taxes until the timber was cut, with the aim of lessening the up-front financial burden of reforestation.

In the brisk post-war years, new markets for by-products led to innovative uses of what was once waste material. The boom in housing construction that was boosting the demand for plywood, invented back in the 1920s,[42] had also opened the door for a brand-new product, particleboard, which could replace solid-sawn lumber in doors, cabinets, countertops, and wall sheathing. Particleboard, essentially a hot-pressed 4-by-8-foot pancake of planer shavings, sawdust, and glue,[43] opened an additional market, besides paper, for wood chips and sawdust that until then had been fueling boilers or otherwise going up in smoke. The wigwam burners that once dotted the landscape of western Oregon began to disappear. Rosboro's burner, built in 1939, was used less and less, and finally the company tore it down in about 1970.

Rosboro's wigwam burner

Rosboro and most of its industry counterparts were pursuing the best conservation and utilization practices they knew. According to a company promotional pamphlet published about 1954:

> Rosboro endeavors to utilize the log completely, either for lumber or wood pulp chips, and 2 carloads of wood chips for kraft paper are produced daily. The remaining sawdust, bark, and refuse not suitable for lumber or pulp are used in our power plant to generate electricity.[44]

In 1951, Rosboro hired its first trained forester, David Burwell. In his thirty-one years with Rosboro, Burwell would take the woods operation to new levels of efficiency, safety, and conservation-minded management. Burwell was a graduate of the School of Forestry at Oregon State College, which trained its foresters and logging engineers in the latest scientific methods of felling trees and reforesting cut-over lands. "Interestingly, at the time Dave went to college," said Paul Cole, "there were many others who trained as foresters that never were able to get jobs using their education, simply because of supply and demand. I suppose the big demand was for loggers, with forestry coming some period after this."

Dave Burwell sizing up a tree

Burwell's first big project was reforesting the Rosboro timberlands on its Finn Rock tree farm. A company promotional pamphlet published about 1954 explains the company's forest-management methods and the reason for them:

> Starting with an abundant supply of good trees, Rosboro seeks to make its timber crop an enduring one through intelligent forest management so that

we may operate into the indefinite future. Our holdings are a dedicated tree farm in which the best known forest management is practiced—protecting the timber from fire, cutting snags..., building access roads for both logging and fire protection, and replanting areas that do not reseed readily.

The pamphlet goes on to describe Rosboro's harvest methods. Its language reflects the postwar spirit of optimism, trust in scientific management, and the widely shared understanding that timber was a crop, just like corn, to be harvested and replanted perpetually:

> Old growth Douglas Fir timber is harvested by an area selection system. Ripe timber is removed in scattered blocks leaving other blocks of live timber to reseed cutover areas. Young Douglas Fir seedlings thrive best on open sunlit ground provided by clear-cutting selected blocks of timber. The remaining areas are reseeded from wind-scattered seeds which come from blocks of trees left standing.[45]

"From the timber perspective, Dave was one of the key people who contributed to the value and longevity of this company," said Gordon Culbertson, who managed Rosboro's timber operations from 1984 until 2001. "A great deal of Rosboro's present wealth and security is a direct result of what Dave did to reforest that Finn Rock tree farm."

Larry Taylor, now Rosboro's contract supervisor, went to work for Dave Burwell, planting trees and washing trucks, when he was a logging-camp brat still in his teens. Taylor's father, Wade Taylor, had worked for Rosboro since the company's earliest days in Oregon. The Taylor family lived in the Finn Rock logging camp, on the south bank of the McKenzie near Blue River. The "camp" was actually a permanent settlement with a couple of dozen houses, a company store, a church, equipment barns, and a large mechanics' shop.

Finn Rock town site

Taylor had been a little boy when Burwell was hired, and he was in awe of the burly forester. "He was about half human and half bull elk," he said. "He'd go charging up through that brush, and nobody could keep up with him. Sometimes we'd be out and it'd get dark, and we'd just lay down under a windfall and go to sleep for the night."

When Taylor was a sophomore at Blue River High School in 1964, Burwell made him the camp's fire warden. A strapping youth, Taylor also helped out on the logging crew whenever it was shorthanded. He worked under the firm hand and eagle eye of Harry Deckert, Burwell's second in command and boss of the logging crew.

"Harry used to come and get me out of Miss Walsager's English class," he recalls. Of course, he adds, she didn't like that very much. "So one day Harry told her, 'I can teach him more English in two hours than you can in two years.'" He doubted that this mollified her. "They didn't care if you were young," he said. "If you could work, you worked."

Logging camp kids, with Larry Taylor on right

He remembers Dave Burwell as a tough taskmaster who demanded loyalty and responsibility. "The first time I ever hauled a Cat, I was sixteen years old. I was washing a truck, and Burwell came out and told me to get into the cab." He ordered Taylor to haul a Caterpillar tractor up to somewhere on the tree farm.

"I said, 'But—but—I don't know how to drive the trailer.'" Burwell replied, "You'll know how by the time you get back."

"I wouldn't trade one minute of his yelling for nothing," Taylor said.

The logging camp has been gone for almost three decades,[46] but Taylor still works the Finn Rock tree farm, as well as Rosboro's several other parcels in Oregon and Washington. "I told myself I'd keep this job until I found something better," he said. "Forty-one years later, I haven't found anything better."

Another of Burwell's early accomplishments was perfecting a logging technique that greatly increased the yield of good timber from the old growth at Finn Rock. The huge trees—"overmature" or "decadent," in the forestry vocabulary of the day—were brittle and apt to break when they slammed into the ground. Because of simple gravity, trees tend to fall downhill, and on a steep slope it is a long way down.

Burwell borrowed a method from Weyerhaeuser Co.[47] for felling trees uphill. He would dispatch a climber to rig a cable to the top of the tree. The cable would run through a pulley positioned uphill from where the tree stood. Then, as the final cut was made by the faller, the yarder operator would give the cable a tug, and the giant tree would sink gently (relatively speaking) into the ground with its top pointed uphill. Uphill felling was slower and trickier than conventional downhill felling, but it kept more of the logs from breaking, and the added value it captured more than made up for the extra time and cost.

Rigged spar on a snowy hillside

Together with his logging boss, Harry Deckert, Burwell developed Rosboro's relogging program. After an old-growth site had been logged, there were always some trees left, smaller but perfectly good. The high-lead equipment used to log the big trees was too massive, and too expensive, to use on the smaller ones. Because it was difficult to log these trees cost-effectively, they were often left as slash. Efficient relogging allowed a company to extract additional value from a site. It also prepared the site for reseeding or replanting, and it reduced the fire hazard by taking off some of the fuel. Deckert and Burwell developed methods that used smaller equipment to do the job efficiently.

Having a supply of smaller logs got the company's leaders, including Paul Cole, to thinking about maximizing their value all through the manufacturing cycle. By the early 1970s, Rosboro would have an entire mill, Mill B, devoted to sawing small logs.

Deckert and Burwell also built an effective safety program for the woods operation. By the time Gordon Culbertson arrived in 1983, "we had gone years and years without an accident in the woods, and our truckers were driving millions of miles without an accident."

Burwell was widely respected throughout the industry, said Rick Fletcher, a forester and professor with Oregon State University Extension. "He had a very high reputation. He was well regarded by many, and very knowledgeable."

The arrival of Burwell and the development of a full-fledged forestry department was an important sign of the values that were coming to dominate Rosboro's corporate philosophy: profitability, careful planning, a long-term vision, and, most importantly, stability. "My father learned a lesson in Arkansas that he didn't want to repeat here," said Paul Cole. "The company there used up all their timber. The company's timber could have been a sustained growing and yield situation. My father wanted to do better. He wanted true sustained yield." Whit Rosborough could have kept his Arkansas operation going indefinitely if he'd been satisfied with two sawmills, Cole said. "But he ruined it by building a third mill" and consequently chewed through his timber supply too fast. "My father had the idea that the yield could and should be sustained. I, too, have that idea."

Branching Out

In the early years of Dave Burwell's tenure, Rosboro had been hiring contract crews, known as "gyppos" in the lingo of the day, to do its logging. In 1954 the company entered the logging business when it bought one of the companies that had been cutting Rosboro timber under contract. The company was Lane Pacific Logging Company at Nimrod, up the McKenzie from

Springfield. Lane Pacific's owner, Fred Rookard, had recently and tragically died, crushed by a piece of loading equipment at Rosboro's log staging area.

Rosboro began negotiations to buy the company from Rookard's widow, Marion Rookard. Also in 1954, Rosboro moved from its cramped offices in the Arcade Building downtown to the Maple School building at 25th and Main, just north of the mill, which the company remodeled into a modern (and, by industry standards, modest) office suite.

Rookard log truck being loaded

In the late spring of 1954, there was a massive lumber strike that shut down operations throughout the Northwest. There had been more than a dozen smaller work stoppages or strike votes since the Depression began to lift in 1936, although patriotic sentiments and wartime production goals produced a temporary hiatus in labor disputes between 1941 and 1945. But now, with the healthy postwar economy, the spike in housing starts, and the high demand for wood products, woodworkers' unions were enjoying the best bargaining position they'd ever had.

The strike took 100,000 workers off the job and paralyzed the Pacific Northwest economy for three months during the summer, the height of the building season. In Lane County alone, six thousand workers walked out.[48] Loggers honored the mill workers' picket line, so both the Rosboro mill and the Lane Pacific Logging Co. were shut down.

The Lane Pacific purchase was completed, and Harry Deckert was put in charge of Rosboro's new logging operation. Deckert was a brusque man with a tough demeanor and inner fortitude to match. After the strike had worn on a few weeks, Deckert told Beuford Cole that he thought he could get a logging crew to go through the picket line. He succeeded, which in hindsight does not surprise Paul Cole. "Actually, this was fairly easy to accomplish because the logging crew was sick of the union, and there had been no union contract signed," he said. "Besides, all the logs from the Finn Rock tree farm were going to the Rosboro sawmill, and the truck drivers went right through the picket lines around the plant."

Paul Cole and Dave Burwell in the woods

After three long months, the strike was settled in the fall of 1954, thanks to exhaustive negotiations of a committee appointed by the governors of Oregon and Washington, Paul Patterson and Arthur Langlie.[49] When the Northwest lumber industry went back to work in September, Rosboro logging operations were already working. Paul Cole believes that "we broke the union with that strike," even though the Rosboro manufacturing operations were and remain unionized. "People need to work," he said. "Loggers are hardy individuals, but they still have to feed their families."

Beuford and Teden Cole on a trip to France

A Brief Retirement

Beuford Cole had been working hard all his life. Now, in his sixties, he was contemplating retirement, a notion that may not have set very well with him. "Paul's father didn't know much more than work," recalled Nonnie Cole. "He didn't know how to have fun or enjoy the fruits of his labors."

He gave it a good try, though. He cut back on his hours at the office, spent more time at home in their garden, and took Teden on a several overseas trips. He seemed to be in fairly good health; he had a little heart trouble but it didn't seem to slow him down any. He may have wondered what his life would be like without Rosboro, how he would fill the years that stretched before him.

He did not get a chance to find out. In the summer of 1958, at sixty-six years old, Beuford Cole died of heart failure, and Paul was abruptly thrust into his father's role. By then he had had more than a decade of mentoring, training, and hands-on practice in all aspects of Rosboro Lumber Company's operations.

Paul Cole was devoted to Rosboro. He had big ideas for the company's future. He was thirty-one years old. He was ready. And in the next decade he would embark on an ambitious program to expand, modernize, and automate the plant—and, in the process, thoroughly transform the company.

Notes

1. Interview with Paul Cole; and Robbins, William G., "Timber and war: an oral history of Coos Bay, Oregon, 1940-1945" (*Journal of the West).*
2. Velasco, Dorothy and Mara Velasco. *Springfield: Between Two Rivers* (Montgomery, AL: Community Communications, Inc., 1999), p. 92.
3. It was Springfield Plywood, started in 1939 by Washington Veneer Co. and acquired by Weyerhaeuser soon afterward. The company was Weyerhaeuser's first venture into plywood sales. *Plywood in Retrospect: Washington Veneer Company* (Plywood Pioneers Assoc., 1972) p. 8. Later, in 1949, Weyerhaeuser built a sawmill and paper mill in Springfield.
4. Ibid., Cole, "Notes on Rosboro history."
5. Minutes of special stockholders' meeting, Rosboro Lumber Company, May 14, 1945.
6. Ibid., stockholders' meeting minutes, May 14, 1945.
7. Minutes of board of directors' annual meeting, Rosboro Timber Company, July 28, 1945.
8. Ibid., annual board meeting RTC, July 28, 1945.
9. "Rosboro Lumber Company History." Undated anonymous memo from company archives, probably written late 1990s.
10. Minutes of special board of directors' meeting, Rosboro Timber Company, March 24, 1947.
11. Smith, Kenneth L. *Sawmill: The Story of Cutting the Last Great Virgin Forest East of the Rockies* (Fayetteville: University of Arkansas Press, 1986), p. 188.
12. Ruth Cole Kainen interview, August 10, 2005.
13. Ibid., Smith, p. 201.
14. Ibid., minutes of annual board of directors' meeting of Rosboro Timber Company, July 28, 1945.
15. Ibid., Kainen interview.
16. Ibid., Kainen interview.
17. Ibid., Smith, pp. 47-49.

18. Paul Cole comments on Chapter 6 draft, July 26, 2006; and ibid., Kainen interview, August 10, 2005.
19. Ibid., Smith, p. 178.
20. State of Oregon Forestry. *Forestry in Oregon: Biennial Report of the State Forester, July 1, 1946 to June 30, 1948* (Salem: State of Oregon Forestry), p. 7.
21. From an annual report from the Oregon-American Lumber Company, quoted in Kamholz, Edward, Jim Blain, and Gregory Kamholz, *The Oregon-American Lumber Company: Ain't No More* (Stanford, CA: Stanford University Press, 2003), p. 243.
22. By way of comparison, in the late 1980s, when the market was good and the spotted owl had not yet hit the headlines, Rosboro was producing about 125 million board feet a year, according to Gordon Culbertson (interview Aug. 31, 2005).
23. "This is Rosboro" (company promotional pamphlet published about 1954), pp. 11 and 15.
24. Township 17 South, Range 4 East, Willamette Meridian; letter from Minot Davis, Weyerhaeuser Timber Co., to Rosboro Lumber Co., Sept. 30, 1941.
25. Letter from B. S. Cole to Cole, Clark and Cunningham, insurance agents, April 18, 1951.
26. Letter from Minot Davis to B. S. Cole, Sept. 30, 1941.
27. Letter from B. S. Cole to Minot Davis, Oct. 6, 1941.
28. Letter from Minot Davis to B. S. Cole, Oct. 8, 1941.
29. Letter from B. S. Cole to Minot Davis, Oct. 9, 1941.
30. To get an idea of how much timberland prices had gone up after World War II, see Kamholz, Blain, and Kamholz, *The Oregon-American Lumber Company: Ain't No More*, p. 243: "In 1951 Oregon-American paid more money for the timber on 210 acres of state lands adjoining the land it acquired from the Salmonberry Timber Company than the company paid for the entire 2,491 acres in that tract in 1940, a more than *tenfold* increase" [italics the authors'].
31. Letter from Minot Davis to B. S. Cole, Dec. 13, 1943.
32. Letter from B. S. Cole to Minot Davis, Dec. 15, 1943.
33. Letter from C. D. Weyerhaeuser to B. S. Cole, June 16, 1948.
34. Letter from B. S. Cole to C. D. Weyerhaeuser, July 9, 1948.
35. Ibid., B. S. Cole to C. D. Weyerhaeuser, July 9, 1948.
36. Letter from B. S. Cole to C. D. Weyerhaeuser, Aug. 21, 1948.
37. Second page of letter from B. S. Cole to unidentified recipient, Jan. 5, 1951.
38. Letter from B. S. Cole to C. D. Weyerhaeuser, Nov. 7, 1951.
39. State of Oregon Forestry. *Forestry in Oregon: Biennal Report of the State Forester to the Governor, July 1, 1946 to June 30, 1948* (Salem: State of Oregon Forestry), p. 76.

40. Ibid., *Forestry in Oregon,* pp. 76-97.
41. Ibid, *Forestry in Oregon,* p. 49.
42. Or rather, it evolved from earlier techniques, some quite ancient, of layering veneer with glue at right-angle patterns. For an interesting thumbnail history, see Michael Thoele, *Bohemia: The Lives and Times of an Oregon Timber Venture* (Salem: Oregon Historical Society Press, 1998), p. 230.
43. Ibid., Thoele, *Bohemia*, p. 330.
44. "This is Rosboro" (Rosboro Lumber Company promotional pamphlet, published about 1954), p. 11.
45. Ibid., "This is Rosboro," p. 5.
46. Bellamy, Ron. "Time's about up for logging camp," Eugene *Register-Guard,* March 18, 1980, p. 1B.
47. "Weyerhaeuser invented the technique, but they never used it," said Paul Cole.
48. Thoele, Michael. *Bohemia: The Lives and Times of an Oregon Timber Venture* (Portland: Oregon Historical Society Press, 1998), pp. 282-5.
49. Ibid., Thoele, pp. 285-6.

4

THE PAUL COLE YEARS BEGIN

"Paul seemed to have the business in his blood."

—Nonnie Cole

PAUL COLE HAD received a good education in the lumber business during his years at Rosboro. He'd had summer jobs in the plant and in the woods. He'd started in sales in 1950, right out of college, and had gradually taken on more responsibility for company projects. One of his early assignments was to develop a program to manufacture wood chips to sell to the pulp and paper industry.

Sawing lumber produced copious volumes of sawdust, hog fuel, wood end trims, and planer shavings. There wasn't much a lumber company could do with this byproduct, and it was a nuisance to get rid of it. Sometimes it was made into hog fuel to power steam engines that ran the mills or sold as wood for home heating, but mostly it went into the wigwam burners that had sprouted like mushrooms across the western Oregon landscape.

Could this waste product be made to turn a profit? Wood chips were needed to make paper (the industry also got its chips from solid logs), but manufacturers like Longview Fibre and Crown Zellerbach were choosy about the quality and species of wood they bought. And they needed a consistent supply.

Longview Fibre, in Longview, Washington, had a 25-year agreement to buy wood chips from Long-Bell Lumber Company at a very low price, $2.50 per unit. "This was a very good arrangement, since the Longview Fibre pulp mill was directly adjacent to Long-Bell's two sawmills," said Paul Cole. By the early 1950s, however, Long-Bell was reaching the end of its timber. Longview

Wood chips in rail car

Fibre general manager Bob Wertheimer approached sawmill owners, aiming to persuade them to install wood chippers in their plants. His goal was to ensure Longview Fibre a future supply of wood chips when Long-Bell had no more.

Paul Cole saw an opportunity. In 1952 he approached Wertheimer and secured an agreement from Longview Fibre to buy all of Rosboro's wood chips, primarily Douglas-fir and hemlock, at $7.00 per unit. Rosboro agreed to install the chipping and railroad-loading equipment. Later Cole added Publishers Paper in West Linn, Oregon, to Rosboro's customer list. Publishers Paper agreed to pay almost double the price that Longview Fibre was paying.

With the growth of this nascent market came the need to produce a consistent supply of high-quality chips. Paul proposed buying some new wood-chipping machinery for the mill. The partners were initially skeptical. "Dad and Tom Watts still had memories of the Depression," Cole said, "and it was hard to get them to put out money for improvements." But he prevailed upon them to spend $50,000 on a chipper and a conveyor. Now that chips were bringing a profit, he argued, it made sense to handle them as efficiently as possible.

Log debarker

Quality was important, too. All wood chips had to be at least 95 percent free of bark. Those sold to Publishers Paper had even tougher standards—"basically, they wanted zero bark," said Paul Cole. Such quality was difficult to achieve when the logs went through the headrig bark and all. In the early fifties, ring debarkers were introduced to the industry, and in 1955, at Paul's instigation, the company installed a 60-inch Nicholson whole-log ring debarker. Rosboro became the fifth mill on the West Coast to install such a machine. Within a few years most of Oregon's sawmills and veneer mills had them.

The development of Rosboro's chip program was a foreshadowing of the sweeping changes Paul Cole would make in his first decade as general manager. When Beuford Cole died in 1958, Paul and the company were both starting to hit their stride. Paul had picked up a broad knowledge of sawmill operations and the lumber industry simply by growing up in his father's family. He had learned much more in his eight years of grooming at Rosboro. "Paul seemed to have the business in his blood," said his wife, Nonnie Cole. "He learned very quickly, but the learning has also been a life-long process."

Paul Cole was also absorbing less-tangible lessons of leadership—careful

Pony rig

Trimmer saws

planning, judicious boldness, fiscal conservatism, and keeping an eye on the competition. These ongoing lessons came from several valuable mentors. One was Eliot Jenkins, president of Booth-Kelly Lumber Company. Jenkins was a colleague and contemporary of Beuford Cole's and a friendly competitor in Springfield. "I tried to get an education and learn from him," said Cole. From Jenkins he learned different lessons than he'd learned from his dad, along with other ways of manufacturing lumber. "For example," he recalled, "Booth Kelly had changed from a strung-out monorail plant to one with crane sheds for handling lumber."

Paul Cole in the early 1960s

For a time George Weyerhaeuser, who was about Paul's age, moved to Eugene to work at the company's large Springfield sawmill, and the two men became friendly. "I believe at some point he was made general manager of both sawmill and Weyerhaeuser logging activities," Cole recalled. "Later when my father died and I became general manager, I was very young. I tried to get around this young age by saying, 'I am the same age as George Weyerhaeuser, and look at the job he has.'"

An important mentor was partner Spencer Collins. Collins was of Paul's father's generation, and after Beuford Cole died he became one of Paul's most ardent supporters.[1] He was a forceful personality. "He ran his own company, Collins and Co., with an iron fist," said Cole. "Young accountants worked for him for not much money, which was typical at the time." Paul would go to Eugene to see Collins and talk business, and their meeting was usually followed by lunch at a restaurant nearby.

Sometimes Collins would drive over to Springfield and drop by the Rosboro office unannounced. He would park in the employees' lot, and, if Paul was absent, prevail on his secretary, Lorene Burt, to let him into Paul's private office. If Paul came in from an errand when Collins was there, Lorene would whisper, "He's in your office!"

One day Paul told Collins diplomatically, "Spencer, I'm always glad to see you, but I have a pretty busy schedule. Next time why don't you call and let me know when you're coming, so I can be ready for you?" After that, he said, Collins would phone and made an appointment.

Collins seemed to feel the need to assert his opinions on the company's management—perhaps a little too much, said Cole. "He just needed to show he was an owner. That can be a struggle if it's not managed properly." Nevertheless, he remained a valuable advisor to Paul, particularly in financial matters. As the company began to accumulate cash in the late 1950s, Paul consulted him about how best to invest it. True to his conservative nature, Collins advised safe, short-term investments. "This turned out to be good advice," said Cole, "and for many years Rosboro made a nice income from the short-term investments, such as money market funds."

Another very important figure for Paul was Bruce Herzinger, one of Collins's top accountants. "He was definitely Number Two in Collins and Company," said Cole. Herzinger balanced Collins's conservatism with an imaginative flair and a tolerance for moderate risk. He advised Paul not to dismiss something as too risky without balancing it against the opportunity that might otherwise be lost—and also to remember that losing money sometimes carries tax advantages.

"Bruce was a wonderful individual, and a very creative accountant," said Cole. "He always had great things to say about me, which I liked." He particularly remembers one of Herzinger's wise sayings: "Paul, if you make a really good deal, you'll never be thanked. If you lose one penny, you'll never be forgiven." Cole said, "That was very good advice—and so true."

Pony deck

Planer

Vertical resaw

Redesigning the Mill

In nearly a decade of observing the workings of the Rosboro sawmill, Paul Cole had concluded that there was much room for improvement. Now he was in a position to do something about it. The sawmill, for instance, was a mess: "I tried to make sense of the lumber flow and I couldn't." The flow from the sawmill to the dry kilns, to the holding sheds, to the planer, to shipping was convoluted and inefficient. For example, some of the lumber was visiting the resaw twice, and there was a lot of difficult, labor-intensive manpower involved in handling of the lumber. "In those days the rule was 'keep the decks to the resaw buried,' and they were, usually about three feet deep,"

Headrig

Edger

said. "Later, the industry learned more production would be realized if we kept the resaw materials as low as possible."

Paul worked with lumber superintendent Jeff Abell to simplify the steps and shorten the process. Here his design mind came to the fore: "I'm mechanical, so I figured out how to change it to a straight-flow operation." Early mechanical layouts were done by George Cleveland, who also supervised the planning mill. Others who worked on the process overhaul were the mill foreman, Lou Northwang, "a colorful foreman," Cole recalled, with a hook in place of a missing hand, and Kermit Alexander, the power plant foreman.

Building the crane shed

One important improvement was a 100-by-500-foot shed with an overhead crane. The crane shed provided a single destination for all the finished lumber and a focal point for improving the flow in the mill. It also made it possible for Rosboro to accumulate a backlog of inventory and fulfill orders faster. Later Paul added a second crane shed for storage of rough dry lumber.[2] Besides better lumber flow, this second crane shed was needed because the large Hyster shed that held rough (and some surfaced) lumber was about to be converted into a plywood plant.

Rosboro's main product line continued to be mixed cars of lumber for home building, and the company was doing very well with it. "We made some fifteen hundred different lumber items in those days," said Paul Cole. Some of these were clear lumber items such as paneling, sidings, and thick, dry, clear lumber. There was some dimension lumber, some shiplap (which

Overhead crane

was used for wall sheathing until sheathing plywood appeared on the scene), and some 4-by-4s, 6-by-6s, and larger items.

Production in 1958 was 200,000 board feet per day,[3] up from 125,000 board feet per day when the company started in 1939. Prices for lumber had leveled out since the boom at the beginning of the decade, but Rosboro was selling everything it was making. Wood chips were selling briskly, too, although the market was getting crowded. The paper manufacturers had found buying wood chips much more cost-effective than chipping logs. The demand had expanded, pulling in new suppliers. "Everybody was making wood chips by then," said Cole.

In 1955 Rosboro moved into another enterprise when it acquired a timber fabrication plant, Miller-Gardner Co. of Monroe, and moved its equipment to Springfield. By 1958 the fabrication plant was producing 30,000 board feet per day of prefabricated roof trusses, arched roof assemblies, bridge and mine timbers, and other custom structures. These were designed from lengths of solid-sawn lumber joined together with mechanical connections, either metal fasteners or a system of precut notches and grooves, or both.

Fabricated timber structures can be as simple as a residential roof truss,

Timber fabricating plant

its inner support members outlining a lazy "W," or as complicated as a bowstring truss or a lamella trussless roof, with its intricate egg-crate tracery. One of the largest lamella roof systems (not a Rosboro product) was used in McArthur Court at the University of Oregon. The advantage of engineered wood structures is that they can cover long spans without the need for additional support. A radial-dome or gothic-style arched roof over a theater, church, or gymnasium keeps the weather out without interrupting the openness of the interior with support posts.

Having a fabricating plant opened a host of new markets and enabled the company to improve the utilization of its raw material. For example, some of the oldest trees on the Finn Rock tree farm contained white speck, a fungal rot. A log from such a tree would not make very good dimensional lumber, but it could be used for engineered mine timbers, which shored up the tunnels until the mine was abandoned. The timbers didn't need to hold up to long-term use, so they could contain more white speck rot than regular studs or dimension lumber.[4]

One big customer for Rosboro's engineered wood products was halfway across the country, in St. Louis. This was a serendipitous connection made by sales manager Val Gardner. Gardner happened to be in St. Louis on Feb. 9, 1959, on a job-related errand, the day after a tornado had hit the city. When

Damaged St. Louis Arena, 1959

Rosboro's lamella trusses at Willamalane Recreation Center in Springfield

he learned about the tornado, he immediately got in touch with authorities and told them Rosboro had just the products they needed to rebuild.

The tornado tore a swath through mid-town St. Louis and ripped a gaping hole in the arched roof of the St. Louis Arena. The roof was a lamella-type clear span of 164 by 450 feet, built in 1929. The tornado also snapped off a 575-foot-tall television tower and damaged the adjacent annex, containing a skating rink and bowling alley.

Gardner took the order on the spot and telephoned it in to Springfield. On Feb. 18 the first carload of Rosboro timbers (12-foot 2-by-10s, cut to size and drilled for bolts) was on the way. It would have been sooner, said Gardner, if a rail car had been available. Two more carloads followed. Other Oregon companies that furnished materials for the rebuilding of St. Louis were the Booth-Kelly and Armstrong mills in Springfield, Kimball Bros. at Dexter, Swanson Bros. at Noti, and Miller Lumber at Monroe.[5]

Getting into Plywood

By now the feverish postwar lumber market had cooled. Things were settling into a more typical pattern, with prices for wood products sometimes up, sometimes down.[6] The 1958 end-of-year price for a Douglas-fir 2-by-4 standard and better was $62 per thousand board feet, about the same as it had been in 1952.[7] The exception was plywood, which was enjoying an upswing.

Modern plywood is an Oregon product. It's generally acknowledged to have been invented in a Portland box factory in 1905, although techniques for slicing wood into thin layers and laying them up in a cross-grained pattern may date back to ancient Egypt.[8] The first modern plywood it was made by Portland Manufacturing Company, a producer of baskets and crates for berries and fruits, as an exhibit for the 1905 Lewis and Clark Exposition in Portland, a grand world's fair celebrating the Northwest's coming of age as a commercially significant region.[9]

The company decided to make display panels featuring Northwest wood species. The manufacture of the panels in March of 1905 foreshadowed the trials and errors the plywood industry was to face in developing a viable commercial product. The factory's crew peeled veneer from logs of ten different species, hung the sheets out on a covered loading dock to dry for two weeks, and finished the drying in a steam-powered kiln. Then they laid up the panels one at a time, hand-painting them with an evil-smelling glue

Power plant

Saw filing room

made from animal blood and starch, and then pressing them between two house jacks.

The Timberman trade magazine's April 1905 issue reported the result:

> [M]anager Carlson of the Portland Manufacturing Co., St. Johns, is preparing a most handsome exhibit for the Lewis & Clark Exposition Fair of his 3-ply veneer work. The samples comprise ten varieties of Oregon wood 40 inches wide and comprise spruce, cottonwood, fir, oak, ash, alder, myrtle, etc. He says he has considerable inquiry for fir veneer glued door panel stock.[10]

This free advertising was no doubt effective, for by October the company was selling veneer panels for trunks and drawer bottoms. By 1912 there were three plywood plants in the Pacific Northwest. By 1929 the Northwest plywood industry was producing 358 million square feet of plywood a year (based on ⅜-inch thickness, the standard index), much of it for door panels and interior paneling for automobiles.

In the mid-1930s plywood surmounted a major technological hurdle with the development of a waterproof glue derived from coal tar. It was a significant improvement over the smelly, unreliable animal-based glues previously used, and it opened the way for the use of plywood in outdoor and marine applications.[11] By the beginning of the 1940s, plywood was going into housing, prefabricated buildings, liners for railroad cars, boats, storage bins and silos for farms,[12] and many other things.

World War II put plywood to wartime use in fighting ships and airplanes, buildings for military bases, crates, huts, barges, bridges, and PT boats.[13] After the war, as with lumber, pent-up demand pushed production and prices ever higher. "...the market soaked up plywood as the scorched desert sand does water."[14] Between 1946 and 1953 plywood production increased from 1.4 billion feet to 3.7 billion feet, and 65 new mills opened in the Northwest. By 1964 there were one hundred fifty-seven plants in operation in the West, producing 10.3 billion square feet per year.[15]

The West Coast dominated the plywood industry, because the best peeler logs came from the big old-growth Douglas-fir trees in the Coast Range and the west slopes of the Cascades. These were used to produce the highest-quality plywood, the sanded and specialty grades.

Rosboro was a relative latecomer to plywood manufacturing. In 1959, when Paul Cole and the other partners began thinking about putting in a plant, there were seventy-two plywood or veneer operations already going in Oregon, most of them ringing the Willamette Valley along the foothills of the western Cascades and eastern Coast Range.[16] Several were close neighbors of Rosboro: Springfield Plywood and Weyerhaeuser each had a plant at Springfield, and Georgia-Pacific had one at Springfield and another at nearby Mohawk. The Eugene-Cottage Grove area had at least ten plywood operations.[17]

True to his methodical nature, Cole embarked on a 6-month investigation of veneer and plywood manufacturing in the spring and summer of 1959. Working with him was Bud Keeney, who had done the engineering on earlier improvements. They contacted equipment manufacturers, got literature on lathes and clippers and hot presses, had long telephone conversations with salesmen and engineers, analyzed price quotes.

"There were a lot of plywood plants being built during this time," said Paul Cole. "The machinery manufacturers or sales companies would call on prospective customers with a large book showing what they had for sale. Companies that we contacted were Prentice, Coe Manufacturing, Moore Dry Kiln Company. Besides their own products they would sell other ma-

Veneer lathe, 1960s

Veneer dryer, 1960s

chinery to round out their lines: veneer lathes, clippers, veneer dryers, hot presses, panel saws, gluing equipment, and other things."

Cole and Keeney traveled the Northwest visiting soon-to-be-rival plywood operations: Cascades Plywood in Lebanon, various plants in Eugene, Roseburg Forest Products' plants in Dillard and Coquille, several Willamette Industries plants, the Weyerhaeuser plant at Springfield, several plants in White City, near Medford, the Woodard Lumber Company plywood plant in Cottage Grove, and others.

Here again Cole's engineering and design skills proved useful. He gathered ideas and made scale models of the likeliest-looking setups, including the one at Woodard's Cottage Grove plant, which he thought would suit Rosboro best. He sought the advice of another competitor, Will Gonyea of Timber Products Company, which had a plywood plant at Medford. Gonyea also ran a sales company called Clear Fir Sales, which sold plywood made by other manufacturers and offered start-up financing. Gonyea recommended that Rosboro set up its plant to produce 10-foot panels, which brought a price premium, in addition to the standard 8-foot panels. "That was good advice," said Cole. "We investigated the difference in equipment cost, and it turned out that a 10-foot plant would cost only $250,000 more."

He knew the plywood plant would be expensive, and he knew how cautious Rosboro's partners were when it came to spending large sums of mon-

Spotting veneer

ey. Paul consulted frequently with Spencer Collins and Tom Watts. They deliberated about costs, forecast potential profits. One day Collins asked him, "Do you think you can keep the cost to $1 million?"

"I told him, 'I'll try,' Cole recalls with a smile. "But I knew all the time it was going to cost $2 million, and it did." Two million dollars was a lot of money in those days. "I tried to keep both Tom Watts and Spencer Collins apprised of what was going on. In a major plant installation such as plywood, money would have to be borrowed, there were some risks, and I needed to have approval from Tom and Spencer."

Three banks were involved in the financing, including Citizens Bank of Oregon, which Paul's father and Spencer Collins had helped establish. "We paid that debt down in two or three years," said Cole. On that subject Paul was as conservative as his colleagues: "We didn't like debt at all."

The veneer plant opened in 1959, and the plywood plant less than a year later. The first superintendent was Merlin Carter; he was soon succeeded by Don Cooley. The plywood operation had one veneer dryer and one hot press. Within a year Paul realized that those weren't enough. He added two new hot presses in 1962, and later a third hot press and another veneer dryer.[18]

High-grade peeler logs were expensive, and there was stiff competition between lumber mills and plywood mills for the best ones. After nearly a de-

The old plywood glue spreader

cade of booming markets, the supply of timber, especially the big, old trees that make the best plywood, was running low on private lands. Increasingly, federal forests were stepping in to fill the gap. In the eight years between 1957 and 1965, harvest of federal timber in Oregon more than doubled, from about 2.3 billion board feet to 5 billion board feet, while the private timber harvest declined by close to the same amount, from 7 billion to about 4.4 billion board feet.[19]

Rosboro by this time was buying considerable quantities of federal timber for both the plywood operation and the lumber side. "When I became general manager," said Paul Cole, "we did almost all our expansion with government timber," which "was a better deal when it first started out [than it is now]. There's always a lot of competition."

On Oct. 12, 1962, a severe windstorm hit the West Coast. The storm took down power lines, damaged buildings, and laid a wide swath of Pacific Northwest forest on the ground.[20] The blowdown "presents both public and private timber managers with the need to move immediately into salvage programs to prevent a bark beetle infestation," according to a column in *Random Lengths*, the West Coast lumber industry's journal of prices and other trends.[21]

As his father had done, Paul monitored markets carefully to determine where to route the best logs at any given moment—whether they'd be more

Plywood hot press

profitable as lumber or as plywood. He worked with the company's forester and timber manager, Dave Burwell, to secure the right types of logs from Rosboro's own tree farm at Finn Rock and from other private suppliers.

By the time the plywood plant was up and running in 1960, the market had flattened a bit, and the company's early returns were disappointing.[22] The price for Douglas-fir ⅜-inch plywood in October of 1962 ranged from $82 to $115 per thousand square feet, depending on grade. "But you don't want to run scared," Cole said. "We're in a cyclical business. We always plan for the downside, because we know there'll always be one."

By 1965, the Rosboro plywood operation was producing 65 million square feet (3/8-inch basis) of plywood per year.[23] A year later the plywood market took a dip, and the company reduced its plywood output by 25 percent and laid off fifty employees. In August of the following year, the Rosboro veneer and plywood plants were down to one shift each.[24]

Another recurring problem that caused plant curtailments during that time was periodic shortages of rail cars. Rosboro had access to one railroad, the Southern Pacific, "and if they were experiencing rail car shortages, we could not ship," said Paul Cole. "If we did not ship, inventories built rapidly, and in some cases, we had to curtail until the car supply improved."

There have been further curtailments of the plywood plant through the years, but the Rosboro plywood operation has survived to the present day, although veneer manufacturing was curtailed indefinitely in November of 2005 because peelable logs had become too expensive, and could be put to more profitable use in the two sawmills.[25]

Moving into Glulam

By the early 1960s, Rosboro was assuming the stature of a fully integrated forest-products operation. The plywood plant, with its drying, gluing, and pressing capabilities, enhanced the company's capability to engineer wood into products more complex than solid-sawn lumber products. The company's plywood and lumber plants together with its timber-fabrication capabilities pushed Rosboro toward the logical next move: engineering the lumber itself. Paul Cole and sales manager Val Gardner began to investigate glue-laminated beams.

Men lifting lumber for the first step in manufacturing a glulam beam

Glulam (the shorthand by which both process and product are now universally known) is made from pieces of lumber of various grades, dried, planed, glued in multiple layers, and joined end-to-end, then pressed, allowed to dry, and cut to different lengths. A glulam timber may be used either as an individual beam or as a component in an engineered structure.

The obvious advantage of glulam is that smaller pieces of lumber can be used to build a big, strong, beautiful beam. A less obvious advantage is that some of those pieces may be of lower quality than the rest, enabling the use of wood that might otherwise be wasted or turned to a lesser use.

Engineering of the beam ensures not only that it is strong enough but that the strength is evenly distributed, with no weak points. Pound for pound, engineers say, a glulam beam is stronger, stiffer, and more uniform than a solid-sawn beam, and just as beautiful.

"We like to think of the beams as a product designed to improve upon nature…as though we were rebuilding the tree," boasts a company brochure published in the 1980s.[26] Another brochure proclaims, "If we can't grow them, we will make them!"[27]

Adjusting clamps for glulam beam

At the time glulam held a tiny share of the building products market. Production nationwide was about 100 million feet per year, according to Cole—a pittance compared to plywood's 9 billion square feet (⅜-inch basis) per year. The market for glulam was not well developed, and sales were highly seasonal—plants would run overtime in the summer and shut down through the winter. In addition, the process linking the order with the manufacturing and final delivery was often hard to control, said Paul Cole. "For a long time, glulam jobs were bid far ahead of time. Even though plants had good estimators for the bids, the actual results when the jobs were produced were generally not as originally estimated." Most of the cost of a glulam beam is the lumber that goes into it, and "guesses about future costs of lumber are generally unreliable."

All the cautions notwithstanding, glulam seemed the coming thing. Val Gardner had a background in timber fabrication; he had come to Rosboro from the Miller-Gardner fabricating company when Rosboro bought it in 1954. Gardner was a strong proponent of putting in a glulam plant. "With [Gardner's] coming in from prefab, and with the lumber we had," Cole said, "it was a very natural thing that we should take our lumber and add value by making glulam beams."

A long load

Cole and Gardner began a research project similar to the one they'd done for the plywood plant. They toured plants in the Northwest and the South, and Cole went to British Columbia to see operations there. Cole hired Ralph Fullbright from the Weyerhaeuser glulam plant at Cottage Grove. Fullbright developed the floor plan for the new plant and selected the machinery needed. The plant would use a horizontal scarf joint to glue the lumber into the lengths needed. There would be two production lines, one on the south side for unusually shaped custom work, and the other on the north side for straighter beams.

The plant opened in 1963. A company pamphlet said, "1963—a busy year. New Glu-Lam Department. Again keeping pace with the latest developments in wood technology for the highest degree of utilization of our timber resources."

Clyde Hughes in the early 1960s

In the winter of 1963, Clyde Hughes, who'd trained as a forester at the University of Montana, learned about Rosboro's venture. He was working at a rival company in Bellingham, Washington, called American Fabricators. Competitors kept pretty close tabs on one another through their standards organization, the American Institute of Timber Construction,[28] and in other, less formal ways. By the spring of 1964, Ralph Fullbright had moved on, and, said Hughes, "I understood that they were looking for a new manager."

Hughes knew Rosboro's reputation for careful planning. He knew the company had large timberland holdings and thus was less vulnerable to being squeezed for logs on the open market. "I saw that their potential was great," he said. "I wanted the opportunity to build a competitive operation."

Hughes got hold of Val Gardner and told him he was interested in the manager's job. Gardner invited Hughes down to meet with him and Paul Cole. In June of 1964 Cole hired Hughes as manager of both the timber fabrication operation and the brand-new glulam plant, which had yet to turn a profit.

"I told them I wanted full responsibility and full authority," Hughes said. "I told them I didn't want someone looking over my shoulder."

In his first few weeks, Hughes sized up the plant and concluded that some changes were needed. He had walked into a situation that is not uncommon in a manufacturing environment: conflict between the sales staff and management about who sets the product's prices and delivery schedules. Management typically complains that salesmen promise too much and quote too low. Salesmen complain that management holds up schedules, takes too long to fix problems, and sets prices too high.

The Rosboro plant in the early 1960s

Hughes saw things as a manager. Glulam beams needed to be custom-priced, not listed on a price sheet by price per board foot, like studs and dimension lumber. "Every [glulam] job needs to be analyzed for materials and time," he said. "That was one of the first things I changed—no more pricing of the product by the sales staff." He also thought the sales department had too much influence over delivery schedule and hence product flow. Within a few months he had taken control of pricing and scheduling, and he had let a number of people go in the process. Cole supported him completely.

Val Gardner went along with the changes, albeit perhaps reluctantly. “He never said anything to me,” said Hughes, “but he was taken aback that I was affecting some of his customers.” In fact, a few customers took their business elsewhere.

Courtesy of Lane County Historical Museum

Southern Pacific train at Springfield station, c. 1962

Next, Hughes addressed the quality of the raw material. Glulam requires lumber that is finished to more exacting specifications than ordinary lumber. For example, the pieces need to be uniformly surfaced to within 0.006 inch from side to side and end to end. Moisture content can be no more than 12 percent, while ordinary kiln-dried lumber can have up to 19 percent moisture. The lumber also has to be of certain grades. “A lot of mills don’t want to be held to that strict a standard,” Hughes said. “So we had a limited source of supply.”

The Rosboro sawmill couldn’t furnish enough lumber of the right specifications, so Hughes bought it from outside suppliers. One of them was Frank

Lumber Co. of Mill City. That company had developed a machine to test its lumber and give it an "e-value," which provided a more accurate measure of strength and stiffness than the standard visual grading system. Using this system, Frank was able to furnish Hughes with lumber of the high specifications he was looking for. Frank Lumber and Rosboro were the first in the industry, said Hughes, to use the e-rated lumber in manufacturing of laminated beams.

Ira Lowe walking from Mill A

Hughes began to build an inventory of laminate grades of 1-inch and 2-inch lumber in the quantities needed to enable an uninterrupted flow of laminated beams, church arches, bowstring trusses, and the other products from the fab and glulam shops. He organized a planning center in the blueprint room where he and his managers could work with the sales staff to set prices and schedules. He hired a new plant superintendent, Al Uskoski, and pulled together an administrative team consisting of estimator John Angvick, engineer Chuck Mitchell, quality-control supervisor Roy Laoli, production control supervisor Ron Kemmer, and secretary Irene Batey. The team was lean by industry standards, Hughes said: "When I'd meet with competitors and we'd talk about our operations, I'd ask how many administrators they had. Rosboro had the fewest administrative staff per thousand board feet [of production] of anybody in the industry."

In its first year of operation Rosboro's glulam plant produced 40 million board feet of product. Rosboro glulam beams began to appear in high-profile places: Willamalane Recreation Center in Springfield, the Valley River Mall in Eugene,[29] the Squaw Valley Olympic Village in Idaho, the Thunderbird Red Lion Hotel in Vancouver, Washington, the power plant at Lane Community College near Eugene, and many others. The company built its new logging bridge across the McKenzie River at Nimrod with its own 5-foot-deep, 80-foot-long glulam girders.[30] The plant could produce beams as big as 14 inches wide by 5 feet deep by 125 feet long.

There were more additions to the other manufacturing plants during that busy year of 1963: a 60-by-100-foot truck loading dock, a new 40-by-160-foot building for storage of green veneer, and a 160-foot expansion to the second crane shed, used for lumber storage, which enlarged it to 75 by 460 feet.

With all this behind him and the glulam plant up and running, Paul Cole barely paused for breath. By the end of 1964 he had built a 240-by-400-square-foot building for plywood storage. The company now had under-cover storage for 4 million square feet of plywood and 8 million board feet of kiln-dried lumber.[31] The company built a 70-by-180-foot addition to the laminating plant, installed a new automatic trimmer for laminated timbers, and began modification of the planing mill, adding an automatic sorting system custom-designed to fit into the existing building.

Enter Rose Burro

Paul had stood behind Clyde Hughes, and indeed behind all his managers, as they made decisions about which people to let go and which to keep. Paul knew that nobody likes firing people—he hated doing it himself. But he knew it sometimes had to be done; that was part of a manager's job. Often he'd discuss a difficult personnel situation with his wife, Nonnie, just as he shared much of the rest of company business with her. The company, after all, was the center of his life, and it tended to dominate family conversations. Paul was often preoccupied with the technical side of the operation, but he knew the more intuitive Nonnie was more in tune with the people side.

They did not always agree. Sometimes Nonnie would urge Paul not to let this or that person go, and he felt he had to. "So I'd go underground," he said. "I wouldn't tell her." Nonnie, for her part, respected Paul's decisions even when she didn't agree with them. "I've seen a lot of people come and go, and some things happened that shouldn't have. Some people may have been placed in positions they weren't ready for. Paul won't tell me when he fires somebody he knows I like." However, she said, her husband and the company "are fair. They do everything they can to accommodate an employee who is not working out."

Rosboro, said Paul Cole, is a demanding employer. Successful employees have to embrace the company's culture and philosophy wholeheartedly. "Some people just don't fit," he said. "And if the company keeps them, [the company is] a fool. The person at the top has to get results."

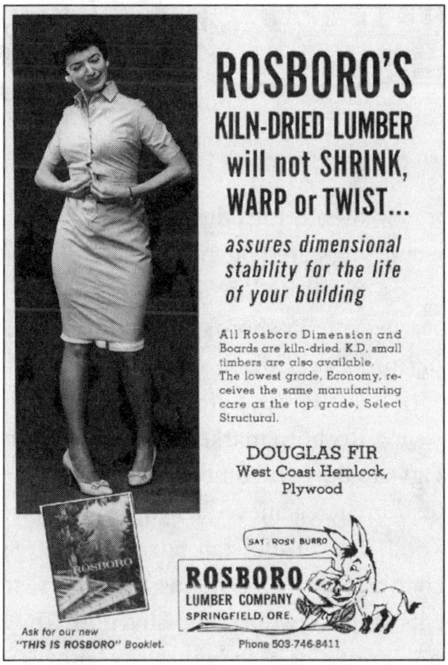

Discreet sex appeal

Clearly, by the mid-1960s, Paul Cole was getting results. The company's products began to appear in advertisements in a host of regional and national industry journals—*Retail Lumberman, Progressive Architecture, Southern Lumber Journal, Crow's Lumber Digest*, and a dozen others. The half-page ads emphasize the quality and innovative character of Rosboro's products and the competence and trustworthiness of the company. "Big enough to do the job," said one ad, "small enough to do it right." Occasionally there a discreet touch of sex appeal: an ad for kiln-dried lumber shows a photo of a shapely woman twisting her torso as she tries to wriggle into a too-small dress. "Rosboro's kiln-dried lumber will not shrink, warp or twist," proclaims the caption.

The winsome burro

By November of 1960 the ads introduced a character familiar to Rosboro's customers today: a cute, big-eared burro sniffing a rose, with a cartoon balloon bearing the words "Say Rose Burro." Rose Burro, the whimsical brainchild of Val Gardner, became Rosboro's face to the world. The creative logo was a welcome departure, Paul Cole thought, from the unimaginative green tree shapes that adorn the letterhead of most lumber companies.

In September of 1967, Rosboro marked the end of an era with the decommissioning of its wigwam burner. Lumber superintendent Ira Lowe told the newspapers that Rosboro had achieved its goal of complete utilization of all its former waste wood. The 79-foot-tall burner, the largest in the Eugene-Springfield area,[32] had become superfluous over the years as the company found markets for its chips, sawdust, and shavings. The *coup de grâce* was the installation of a "hammer hog" to pound any leftover waste material into hogged fuel. Rosboro used about half the fuel to fire its own plants and sold the rest.[33]

The burner stood idle until about 1970, and then it was torn down and sold for scrap. Its dismantling marked the passing of a unique period in western Oregon, a time when small-town lumber and plywood mills were the backbone of the economy and culture, their dominance symbolized by the ubiquitous presence of woodsmoke in the air.

The world was changing. The first Earth Day was on the horizon. Within the next two decades, players in the lumber industry would feel pressure from many quarters: a globalizing economy, uncertainties in the supply of raw material, and—most dismaying of all—an abrupt about-face in the way Oregonians regarded their magnificent forests and the people who worked in them.

Industry leaders were used to being the good guys. They gave people jobs, they contributed to the community, they were good citizens. Weren't they? They weren't ready to play the villains.

Notes

1. More supportive, maybe, than Beuford Cole had been. "My father was the kind of person who never gave credit," said Ruth Kainen. "He never complimented Paul and told him what a good job he was doing. And my brother is brilliant." Interview with Ruth Cole Kainen, Aug. 10, 2005.
2. "This is Rosboro" (company promotional pamphlet published about 1964).
3. "Rosboro Timber Fabrication" (company pamphlet published in 1958).
4. Paul Cole note, Jan. 5, 2006.
5. "Six lumber mills join forces to speed repair of arena's tornado damage" (*Wood Construction & Building Materialist*, June 1959), p. 26.
6. Thoele, Michael. *Bohemia: The Lives and Times of an Oregon Timber Venture* (Portland, OR: Oregon Historical Society Press, 1998), p. 305.
7. Joe Heitz, associate editor of *Random Lengths*, personal communication Jan. 23, 2005. Price data from *Random Lengths* archives, comparing Douglas-fir #2-and-better 2-by-4s in 1952 with Douglas-fir standard-and-better 2-by-4s in 1958.
8. Cour, Robert M. *The Plywood Age: A History of the Fir Plywood Industry's First Fifty Years* (Portland, OR: Douglas Fir Plywood Association, 1955), p. 3.
9. Spence, Mark. "Soyaapo and the remaking of Lewis and Clark" (*Oregon Historical Quarterly* 105(3), Fall 2004).
10. Ibid., Cour, pp. 13-16.
11. Ibid., Cour, pp. 91-2.
12. Ibid., Cour, pp. 115-16.
13. Ibid., Cour, pp. 134-38.
14. Ibid., Cour, pp. 139-40.
15. McKeever, David B., and Gary W. Meyer. *The Softwood Plywood Industry in the United States, 1965-82* (Madison, WI: Resource Bulletin FPL 13, USDA Forest Service Forest Products Laboratory, 1984), pp. 5-6. In the mid-1960s, operators

in the South began to make inroads into the industry, producing mostly lower-grade sheathing plywood from southern pine species.

16. Ibid., McKeever and Meyer, pp. 4-6 and 20.
17. Ibid., McKeever and Meyer, pp. 4-6.
18. "This is Rosboro" (company pamphlet published about 1964).
19. Beuter, John. 1998. *Legacy and Promise: Oregon's Forests and Wood Products Industry* (Portland: Oregon Business Council and Oregon Forest Resources Institute).
20. Estimates of the amount of timber blown down range from 11 billion board feet (http://en.wikipedia.org/wiki/Columbus_Day_Storm) to 15 billion (http://www.noaanews.noaa.gov/stories/s1054.htm) to 16-17 billion (http://asae.frymulti.com/abstract.asp?aid=18056&t=2).
21. "Through a Knothole," column published Oct. 19, 1962, reprinted in *Milestones on a Bumpy Road: Collected Articles from Random Lengths* (Eugene, OR: Random Lengths Publications, 2002), p. 13.
22. Interview with Paul Cole, July 8, 2005.
23. Ibid., McKeever and Meyer, p. 15.
24. "Rosboro Plans 50% Shutdown." Article in Springfield News [?], apparently June 1967. (Clipping in company scrapbook; date and name of newspaper not included.)
25. Random Lengths online edition. "Woodwire: Curtailment Watch." (http://www.randomlengths.com/base.asp?s1=Daily_WoodWire&s2=Curtailment_Watch&s3=Panel_Mills&pub=list, consulted Jan. 23, 2006.
26. "A Wholesaler's Guide to Q-Beams." Company pamphlet; undated but mentions overhaul of "Big Log" mill in 1980.
27. Ibid., "This is Rosboro," about 1964.
28. AITC was formed in 1952 by several glulam manufacturers to develop standards for structural laminated timbers. Rosboro is a charter member.
29. These two examples came from company archives; the next two came from Clyde Hughes; the fifth came from a newspaper story of Aug. 21, 1967.
30. "This is Rosboro," (company promotional brochure published about 1964).
31. Ibid., "This is Rosboro," 1964.
32. "Area's Largest Mill Burner Ends Service" (Springfield *News*, Sept. 28, 1967); "Wigwam Burners: These Closures Clear Air" (Eugene *Register-Guard*, Oct. 14, 1967). The *News* says the burner was 70 feet tall.
33. Ibid., Eugene *Register-Guard*, Oct. 14, 1967.

5

FROM SAWDUST TO SILICON

"This company has always been on a path of continuing improvement."

—*Paul Cole*

In 1965, Rosboro Lumber Company turned twenty-five. The company threw a party at the Lane County Fairgrounds, and eight hundred people came to eat a steak dinner, dance, and listen to speeches from company executives. The owner-partners were all there, including Rosboro's grande dame, Anna Rosborough, widow of the founder.

Paul Cole has fond memories of Anna Rosborough. "Anna was a wonderful person, extremely friendly, great memory, had a nice way to say everything," he said. "She was a truly cultured lady. She went out of her way to say what a good job she felt I had done in running the company." Cole was in the habit of telephoning Anna Rosborough to give her some really good news, "such as when we had reached the $1 million income level on the previous month's operations. As usual, she thought this was great."

Val Gardner, Rosboro's ebullient sales manager, emceed the event and a few weeks later noted it, with uncharacteristic modesty, in a newsletter article: "With many companies celebrating 50-year birthdays, and even centennials, we won't claim that a mere 25th anniversary is any great event." Nevertheless, he was clearly proud that the company had made it this far. "Many independent mills [are] falling victim to closures or mergers, but Rosboro continues to grow with total employees around 550. Total lumber production this year will be around 57 million bd. ft., and plywood production around 65 million sq. ft. on a ⅜" basis."[1]

When Rosboro Lumber Company decommissioned its wigwam burner in 1967, it was without a shred of nostalgia. The future, not the past, was the land of promise. Rosboro was a company in its prime and on the rise, jostling for leadership among the hundreds of companies that had entered the race during and after the Second World War.

Its single-minded quest for competitiveness would lead Rosboro on a path of major changes in the next decade. On the technical side, there would be extensive, expensive upgrading of the sawmill and new equipment for the plywood and glue-laminated beam plants. On the people side, there would be a gradual but decisive shift toward a systematic, team-oriented management structure. Other strong personalities, besides those of Whit Rosborough and Beuford and Paul Cole, would begin to leave their mark on the company.

Group portrait at twenty-fifth anniversary banquet

Rosboro was still in the mixed-car trade, manufacturing dozens of different products: green and dry Douglas-fir and hemlock studs; green and dry fir and hemlock dimension lumber in a dozen sizes;[2] Douglas-fir timbers in thicknesses of 6 and 10 inches and lengths up to 32 feet;[3] panels up to 5 by 10 feet and textured as shiplap, tongue-and-groove, T1-11, and many other shapes; sanded plywood in two grades and six thicknesses;[4] veneer (mostly used in Rosboro's own plywood plant); timber beams; custom-fabricated timber assemblies; and custom glue-laminated timbers in dimensions up to 5 by 14 inches and lengths as long as 125 feet.[5]

It was a lot of products for a smallish company. The mixed-car trade represented a sort of general-store approach to wood-products marketing: a company would position itself to furnish a little bit of everything a builder might need to build a house, everything a retail-yard operator might need to fill a warehouse. It took a lot of manpower to produce and keep track all these products, some of which were only marginally profitable.

Market pressures through the mid-1960s to mid-'70s would combine with improvements in process control to push many lumber companies (and other types of businesses, too), away from this general-store approach. Companies began to have the means and the motivation to move toward a more specialized product line, focusing on what the company did best and what would bring the most profit.

Lumber and plywood markets in the early to mid-1960s saw a few sharp peaks and troughs, but the average price of these commodities rose slowly during the early 1960s.[6] Log prices, in contrast, rose by more than 60 percent between 1963 and 1968.[7] An export market for logs had opened in Japan, which boosted demand for West Coast hemlock and Douglas-fir logs. And in October of 1962 a freak typhoon hit the West Coast and downed millions of trees, containing between 11 and 17 billion board feet of timber. The Columbus Day Storm did not by itself trigger the Japanese export trade,[8] but it did make available many logs that might otherwise have stayed on the stump for a while, and it eased life for log-short mills for the next few years.[9]

The Japanese log market was a boon for those operators who had timber of their own to sell, because they could then bid on Forest Service timber to keep their own mills running (federal logs were banned from export in 1968[10]). It was bad news for the smaller operators and those with no forestlands, because it priced some of them out of the log market. The Japanese export market would play a major role in the West Coast timber market until the early 1980s. It was an important factor in the industry's mid-1960s shift from sawing mostly private timber to sawing mostly government timber, which made the industry dependent on what turned out to be very chancy source of raw material.

Aerial photo of the Rosboro mill, c. 1964

There were more market ups and downs in the late 1960s and early 1970s, a time when inflation was building throughout the U.S. economy. A price rise early in 1971 triggered price controls that continued through 1973.[11] Adding to the pressures were a new lumber-grading standard published by the

Commerce Department in 1970[12] and an intermittently severe shortage of railroad cars throughout the late '60s.[13]

Rosboro would emerge in the mid-1970s with a brand-new stud mill, a new glulam plant, a pared-down product line, and a policy of relentless improvement. By now, also, Rosboro was taking its place as an economic mainstay in Lane County, having become one of the ten largest taxpaying companies in the county.[14]

More Improvements in the Plants

Rosboro had started its glulam operation by making custom beams in response to special orders from customers. By 1966 the company had added a line of stock beams, six sizes in a 3¼-inch thickness and six more in a 5¼-inch thickness, up to 60 feet long.[15] At the end of that year Cole and Hughes decided to buy a new finger-joining machine, which was able to make strong joints with shorter pieces of wood. The horizontal scarf joints had to be 10 inches long, while the vertical finger joint required only 1¼ inches of wood. The machine was more efficient in use of raw material and also required less manpower to operate. Rosboro became the first glulam producer in the industry to use a "continuous-flow" finger joint line, where individual laminations were cut to predetermined lengths automatically as they sped down the line.

The next improvement was a radio-frequency press, installed in 1975. The new press dried the glue and cured the beam much faster than the former method of clamping the beam and letting it cure—a matter of minutes rather than hours.[16] Under the old method, because the glues used at the time were slow to dry, the beams had to remain under pressure for a long time, 8 hours on the average, longer if the weather were cold or wet. Heaters of various kinds were used to speed the drying.

Besides being much faster, the new press made beams continually, without any stops. Costs were lowered because it required only two operators, rather than the nine to eleven required by the manual-form procedure, and it used one-third as much glue, an expensive item. Before long Rosboro was producing 100,000 board feet of laminated beams per day in an almost continuous-flow operation.[17] The pieces of lumber were first finger-jointed, then fed through the radio-frequency glue-drying tunnel. The next step was to put these long pieces into stacks the same size as the beam to be produced.

Next, the lumber flowed beneath glue spreaders, and then the stacks were fed lengthwise into the radio-frequency press and made into glulam beams.

Soon after the finger joiner was installed, Clyde Hughes had the finger-joined lumber tested by an independent laboratory. The testing consisted of bending the lumber until it broke at the weakest point. In most cases the finger joints held firm even after the solid wood broke.[18] "The new finger joint, in my opinion, is the greatest thing that has happened at Rosboro since I have been here," enthused Val Gardner in the *News Burro*, his semi-monthly newsletter to wholesalers. "The new product possibilities are almost endless."

Finger joiner

Frequently, Paul Cole would walk the quarter-mile from the office to the laminating plant to see what was going on. He listened to Hughes's reports with interest, but typically ended the conversation with the question, "When are we going to start making some money?"

"It won't be long," Hughes promised. As it turned out, however, the glulam plant took a while to pay for itself and even longer to turn a profit. "We naively thought that there was a close connection between lumber manufacturing and glulam beams," said Paul Cole, "but the truth is, they are completely different businesses." The glulam market was too much controlled by the customers, he said, production was inefficient because of the many types of products available, and the competition could be ruinous.[19] In hindsight, "we should have researched it better." Eventually, however, the glulam operation proved to be one of the company's most consistently profitable product lines.

By 1967 Rosboro, in addition to purchasing glulam beam lumber from outside suppliers, was using some of its lumber for stock for laminated timbers, and was offering little or none to wholesalers.[20] Utility-grade studs were perfectly good for most framing uses, Gardner told his wholesalers. "Our contention [is] that Std&Btr grades of Fir are just simply too good a lumber to use for ordinary joist and rafter construction. This is the reason that we use all of this grade for lam stock where its higher strength characteristics can be used to full value." The decision to divert the better grades of lumber into glulam is one example of Rosboro's growing effort to turn each product to its most profitable use.

Improving the Sawmill

Jeff Abell was the lumber superintendent during Paul Cole's early years as general manager. Abell retired in 1965[21] and was succeeded by Bernard Lockman. Then, in the summer of 1966, "a fellow named Ira Lowe walked into the office looking for a job," said Cole. It was a serendipitous occurrence for both men, and for the company.

Lowe had been superintendent of Weyerhaeuser's Everett, Washington, sawmill until a few months before. Then, in a management reorganization, he had been bumped down to head saw filer, and decided he didn't like that very much. Through a contact with a saw salesman who serviced both companies, he heard that Rosboro might be looking for a new sawmill superintendent. In the summer of 1966 he traveled to Springfield and met with Paul Cole. There were no positions open then, but Cole could see that Lowe was an energetic, can-do kind of guy, and he was definitely interested.[22]

Soon after the visit, Cole telephoned Lowe in Everett. Bernard Lockman was moving on. Was Lowe interested in the superintendent's job? Lowe was. He started work at Rosboro in September, and thus began a warm and productive collegial friendship that would influence Rosboro's development in numerous ways—and that would end abruptly twenty-six years later.[23]

Despite a lengthy tenure at Weyerhaeuser, Ira Lowe "did not fit the Weyerhaeuser mold," in Paul Cole's judgment. "Which means that he liked to do something more than just take orders. He had ideas of his own, and he was pretty aggressive." Val Gardner described Lowe thus: "Ira has a built-in generator for producing ideas, suggestions, and actions for cost reductions and product improvements." In his characteristic breezy style, Gardner added, "He keeps me hustling to keep up with him, but it sure makes the job a lot more interesting."[24]

Ira Lowe

Paul Cole had already begun many innovations in the sawmill. Lowe's arrival accelerated this process, and the two of them talked and worked together on the changes they thought were needed in the constant quest for higher volumes, lower costs, better production per man-hour, and the most profit for each product. Lowe sized up the sawmill and agreed with Cole that the machinery, process flow, and product line all needed to be upgraded and streamlined. Their first innovation was to investigate thin-kerf saws, which had the potential of greatly improving overrun on the logs.

They went to see an innovative entrepreneur in Ukiah, California, named Al Thrasher. Thrasher had devised and patented a sawing system with specially designed saw guides and very narrow circular saws. The system greatly improved wood utilization by reducing saw kerf—the amount of wood chewed up by the saw's blade. Thrasher's company offered licensing agreements to sawmills and charged a royalty for the use of its design.

In the end the company did not use any of Thrasher's equipment. However, Cole and Lowe borrowed some of his ideas on reducing saw kerf on the company's bandsaw headrig and edgers. "With bandsaws," said Cole, "kerf was reduced by adding more strain [tension] on the bandmills, better speed controls, different feed speeds, or replacing existing band mills with high-strain units."

Cants going to pony rig

Being a saw filer, Lowe found that tinkering with saws "was right down my alley." He and Cole took out the old pony rig and put in an automated sawing system made by Maxi Mill, an Albany, Oregon company. The system consisted of an overhead end-dogging carriage combined with automated control of the log as it passed through the saws.[25] The logs were scanned and positioned prior to being dogged onto the overhead carriage.

The system, one of the earliest to use computer technology in sawmill machinery, was an early Maxi Mill prototype, so there were several bugs to work out. As they were working on the sawing systems, Cole and Lowe also improved the process flow within the mill, put in an automatic lumber sorter, and added a section to the mill building for new twin vertical resaws.

After considerable revision and tinkering, Lowe and the sawmill crew got the new systems working smoothly. Rosboro started seeing gains in overrun—meaning the amount of board footage in product gained for every board foot of log scale. Overrun is a standard shorthand for how much product is squeezed out of a particular log.[26]

These improvements got Ira Lowe and Paul Cole thinking about the dimensions of logs coming into the mill. Big old-growth logs were at a premium; they were prized by high-grade sawmills that had access to export markets and by a few specialty plywood plants, and they were also sought after for the Japanese export market. Smaller second-growth logs, up to 12 inches in diameter, were easier to get, and they were cheaper by as much as $100 per board foot.

"We wanted to do what we do best, and do it well," said Lowe. "We had many small logs coming into the big mill, and [after the improvements] we saw that we could do very well with small logs. So the company started selling the biggest logs and bringing the smaller ones into the mill. That way we made money selling logs and processing logs, both."

The profitability of milling small logs got them thinking even further. What if the company could have a specialized process for just the small logs? "We realized," said Lowe, "that we needed a little stud mill."

Before plunging into this newest venture, Cole and Lowe did their customary careful research. They traveled to many mills in the Northwest and to Scandinavia, where an overhead carriage system known as the Ackerfeldt system was being used by some sawmills. Knowing that a small-log mill would have to be well coordinated with the logging operation, they consulted with Dave Burwell on how the log supply was likely to shape up.[27] Burwell's predictions proved conservative. "Our forester's forecast fell far short of the actual supply," Cole said.

With this project, Paul Cole's keen engineering sense combined fruitfully with Lowe's inventive streak—Lowe had been a Seabee in the Navy during World War II. Together they designed a small-log stud mill and started building it inside the existing planing mill building, which was standing nearly empty.

When Paul Cole first arrived in 1950, the company was planing some of

Log debarker and Mill A

its dimension lumber in that building and using the rest of the space for making moldings, siding, and other finished products. But that was almost twenty years ago, and now Rosboro's product mix was considerably thinner. The company was making fewer dimensions of lumber, and the planers in the big mill building were handling all of it. The millwork products had been phased out altogether. Thus there was a building to house the new stud mill.

The first sawing system Cole and Lowe installed was a chain with a pusher that pushed the logs through the quad bandmills. This worked very well, Cole said. "Then Ira had another idea, a double carriage, one carriage over the other carriage, so you could be passing one carriage through the saw while the other was getting into position. It wasn't constant motion, but it was pretty good."

That design, aided by periodic upgrades in computerized process control, is still the heart of the small-log mill, which became known as Mill B. At first set up to produce studs 8, 9, and 10 feet long, Mill B would later be retooled to produce 16-foot 2-by-4s, a very popular item.[28]

These upgrades were not cheap, but in Paul Cole's estimation they were critical in keeping the company competitive. "We spent an average of $2

Stud mill for sawing cores, 1960s-1970s

million every year for many years updating, modernizing, and expanding our plants," he said. "This company has always been on a path of continuous improvement."

By now Cole had been leading the company for almost a decade, and he'd built up a good track record. The result was that the owner-partners were willing to trust him with multi-million-dollar decisions. "The partners say to me, 'We expect management to run the company,'" said Paul Cole. "A lot of owners [in other companies] say, 'Well, we'd like management to run the company, but we'd sort of like to tell them what to do.'" Ruth Cole Kainen, Paul's sister and, like him, a second-generation partner-owner, calls her brother "brilliant" in his ability to envision the company's future course, invest aggressively in improvements, and hire good people.

During this time, also, the ownership of Rosboro Lumber Company was beginning to branch into a complicated family tree. The Beuford Cole share passed to Paul and his two sisters, Elizabeth Butler and Ruth Kainen, when their father died in 1958. Paul Cole had purchased a percentage of the Vernon Williams interest in 1951, when Williams left the company. At the same time Beuford and Teden Cole had sold the same percentage each to Ruth and Elizabeth Cole.

The Spencer Collins share passed to an annuity trust for the benefit of three colleges, and Collins's successor on the company executive committee was his stepson, Chuck Slemons. After Slemons died, Collins's own son, Creighton Collins, succeeded to the executive committee. After about a year and a half Paul Cole negotiated a buyout of the Collins interest, and it was split proportionately among the other owners according to percentage of ownership.

Tom and Martha Watts still owned their portion, but they were childless, and by now Tom was ailing with Parkinson's disease. Within a few years his portion would pass to his nephew, James Foster Watts, Jr., and nieces Roberta Watts Case and Susie Watts Margolin, the children of Tom's brother James Foster Watts. Eventually Martha's share would pass to William Hubert ("Hub") Lindsey Jr., son of her twin sister Mary Lindsey; and Emmett Pratt and Nelle Pratt Scott, son and daughter of her other sister Elizabeth Pratt.[29] These transfers happened over a period of years and in a variety of complicated ways involving trusts, S corporations, and other legal structures and arrangements.

End of the High-lead Era

Operations in the woods were also changing. Rosboro's loggers, under the direction of the company's forester Dave Burwell, perfected the uphill-felling technique Burwell had adapted, and the company was leading its peers in saving trees from breakage.[30] In the spring of 1968 Rosboro announced that it would cease its two high-lead operations at the Finn Rock tree farm and lay off twenty-five woods employees. Henceforth the logging would be done by contract crews, which, with their more modern equipment, were cheaper and faster, and the operations caused less breakage of the logs.[31]

Burwell told the Springfield *News* that the company couldn't afford to buy new woods equipment just now. "[B]y making the change at this time," he said, "just when the woods are beginning to open up again we're hopeful [the laid-off men] will be able to go to work in other logging operations." The news story concluded on a nostalgic note: "But the high climber, long an indispensable man in the woods, will soon be only a professional entertainer to be seen topping spar trees in exhibitions at logging shows…"

Rosboro was coming to the end of its old-growth timber at Finn Rock, anyway.[32] The opening of the National Forests and BLM lands for logging helped Rosboro and many other companies stay in business as much of Oregon's privately owned timber was depleted through the 1950s and '60s. By the early to mid-1970s the company was logging as little of its own timber as it could get by with, between 17 and 20 million board feet per year. About 80 percent of its timber came from National Forest or BLM lands, and a small amount came from private timber purchases.[33]

Strong-willed Managers

Gene Scherffius, hired in 1969 as assistant office manager, was promoted to office manager after Al Maupin became seriously ill. Trained in business management and forestry, he was also was put in charge of the woods operation. This was part of a move by Paul Cole to bring better planning and focus to the timber side of the business. Scherffius assumed other duties as well, including that of controller, "although I was never given the title."[34]

Also about this time Cole, who had been dealing with the company's managers individually, pulled them together into something resembling a team. "The managers still had authority and responsibility to run their divisions," Cole said, "but by meeting as a group, we were able to develop more ideas."

Each manager had a forceful personality and strong opinions on how things ought to be done. Scherffius was a decisive executive, a do-it-now sort of person, and he was ambitious.[35] In Scherffius Cole saw a man who could bring order and regularity to the timber operation. "We had a situation with Dave Burwell, because he did not like to follow plans; he liked to make plans as he went along," Cole said. "We had to have more structure than that. So I gave Gene the opportunity to coordinate the timber operation, because I could not get planning out of Dave Burwell." In 1969 Scherffius was promoted to timber manager. He was a man of regular habits and a delegator, not a hands-on manager. He would arrive early in the morning to meet with his staff, and he would often leave early in the afternoon, which, he said, did not sit well with Paul Cole.[36] Eventually these traits began to tax Cole's patience and would become a source of friction between them.

Clyde Hughes had definite ideas of how to run a glulam operation. When he interviewed for the job in the fall of 1963, he told Paul Cole he didn't want it unless he could have complete control. Hired the following spring, he wasted no time in firing the plant superintendent, assembling his own team, and imposing his own process. Soon Hughes was firmly established in the glulam plant, about a quarter-mile from the office on 25th and Main. There he managed his own secretary and conducted his own sales. Hughes's seeming independence from the larger company's operations would prove a sore point with some of the other managers, and, ultimately, with Paul Cole.

Bryce Weissert was hired as controller, and he became a very good one, said Cole. "I taught him how to summarize a financial report or reports on one page of paper, and he became good at this." He adds that, for a financial type, Weissert "was also a lot of fun."

Then Weissert asked for a job in production, so he was put in charge of managing the glulam plant after Clyde Hughes' departure in 1980. His personality, background, and skills did not translate well to the demands of managing a complicated operation like the glulam plant, in Cole's estimation. "Financial skill was important here, and he was good at developing and analyzing figures. He was not so good at dealing with people or with sales and production problems." Within a couple of years Weissert was asked to leave.

Ira Lowe was by most accounts the strongest personality on the team. He was decisive or arbitrary; a visionary or an autocrat, depending on whom you talked to. He had a rare focus, enabling him to reduce problems to their common denominator and pay attention to what was important.[37] He was a gifted mechanic and inventor, able to craft ingenious solutions to vexing process problems—a gift that would prove less useful with time, as the mill's equipment became more sophisticated and more computerized. By most accounts he could be overbearing with his employees.[38] By the time he retired in 1993, Lowe had risen to the position of operations manager, in charge all the company's manufacturing processes—lumber, plywood, veneer, and glulam. But his last few years with the company were troubled, and his retirement would not be entirely voluntary.

Val Gardner was not on the management team, but he was Rosboro's public face, a cheerful and tireless promoter of the company and its products. It was his brainstorm that led to the company's logo, the cute little burro with the rose. In April of 1965, following in that whimsical theme, Gardner created the *News Burro*, a semi-monthly newsletter for the company's wholesale customers. It contained two dense, back-to-back pages of text, every word written by Gardner.

A main purpose of the newsletter was to tell customers what the company did and did not have in stock: "We are pretty badly sold out of almost all dimension items both Fir and Hemlock. We are fairly comfortable on most items of Fir uppers. We have some Hemlock D uppers, but C and Btr is oversold."[39] Sometimes he would press a point in his chatty but firm manner: "If there has been a day go by that we haven't had to tell somebody that we do not make any 4 x 8 Sheathing, I have never seen that day yet. That means exactly what it says. We do not make any at all."

Val Gardner

He would use the rest of the newsletter to brag about the company, tell jokes, bemoan the market, recount fishing trips, and exhort his wholesalers on to greater sales glory. "I have said it before, and I make no bones about saying it again. We manufacturers need you wholesalers, not only to peddle our merchandise, but to help us promote it." Sometimes he nagged them: "He who whispers down a well about the goods he has to sell will never reap the silver dollars like he who shows them round and hollers."[40]

Occasionally he used the *News Burro* as a bully pulpit. In the March 12, 1968 edition, after his usual report on the availability of the company's various products, Gardner urged support of a bond measure that would finance a new forestry building at Oregon State University. "It is our opinion that the state of Oregon, with the importance of the forest products to its economy, should have the greatest forestry school of any school in the country," he declared. "Now, I solicit you fellows in the forest products industry, whatever phase it may be, to get out and campaign with your neighbors and friends to vote in favor of this bond measure."[41]

Gardner had been at Rosboro longer than anyone in management except Paul Cole. He had grown comfortable with the multitude of products Rosboro was making in the early years, and he'd built up long-standing and cordial relationships with wholesalers and commercial builders who had come to rely on Rosboro as a one-stop shop. He was not so comfortable with constantly hustling new customers, which became necessary as the company pared down its product line. That would eventually make life at Rosboro difficult for him.[42]

Turmoil Outside the Gates

Outside Rosboro's gates, the conservative 1950s were giving way to turbulence, rebellion, protest, sex, drugs, and rock 'n' roll. The Baby Boom, seventy million strong, was growing up, entering college, and confronting military service in a variety of ways, from enlisting in the military to complying with the draft board to burning draft cards to fleeing to Canada. The 1960s brought in a ferment of social change: the civil-rights movement, the feminist movement, and—most pertinently for the timber industry—the environmental movement. The decade ushered in revolutionary ways of thinking that changed America's culture irrevocably. There was no turning back.

The closing of the decade brought violent race riots in Newark, Chicago, Los Angeles, Detroit, and New York. In 1968, two revered and controversial figures, Robert F. Kennedy and Martin Luther King, Jr. were assassinated. Looming over everything like a malignant cloud was the Vietnam War. Its images poured into living rooms nightly, vivid in the living color of television. The war opened deep fault lines in the nation's political landscape. College campuses, including the University of Oregon at Eugene, erupted in protest.

The American environmental movement sprang into the public consciousness with Earth Day, April 22, 1970, although it had its roots in the nineteenth century, as the nation ended its task of taming the wilderness and began to appreciate nature's aesthetic and recreational qualities. The 1960 Multiple Use–Sustained Yield Act put recreation on a par with other land uses in federal forests, and the Forest Service and BLM developed campgrounds, trails, and other facilities to accommodate the growing number of campers, hikers, anglers, and skiers heading into the back country.

The Willamette National Forest, right in Springfield's backyard (and a major producer of Forest Service timber) was also a prime recreational destination. The area is huge—1.7 million acres covering three major river drainages along the west flank of the Cascades,[43] and taking in parts of five mountain peaks, six counties, three congressional districts (out of Oregon's five), and four major highways. Even as sawmills in western Oregon were becoming more dependent on the Willamette's timber, the public was becoming fonder of its sparkling lakes, verdant meadows, and breathtaking vistas. In 1953, 115,000 campers and anglers visited the forest; three years later they numbered 190,000. In 1966 the forest was host to 225,000 campers alone.[44]

In 1964, Congress passed the Wilderness Act, setting aside 9.1 million acres on national forests as federally protected wilderness. The goal of the legislation was to preserve these lands in their current undeveloped state as "an area where the earth and its community of life are untrammeled by man, where man himself is a visitor who does not remain."

Even though a fair amount of timber was now flowing out of the Willamette National Forest, there still were many thousands of undeveloped acres. Major controversies arose over which land, and how much, ought to be set aside as wilderness. The legendary battle over the French Pete area between 1968 and 1978 pitted the timber industry and other commodity interests against an increasingly vocal environmental movement, and it became an emblem of what hardened into a long-standing ideological polarization between these factions.

The French Pete issue arose with a proposed timber sale on a parcel of forest adjacent to the Three Sisters Wilderness. The 53,000-acre area, known as French Pete, or Olallie Ridge, had been proposed for inclusion in the Three Sisters Wilderness when it was designated in 1957, but in the end it had been excluded, to the disappointment of environmental groups.

In 1968 the Forest Service drew up a timber sale in the French Pete area. A Eugene-based group called the Save French Pete Committee rallied local opinion against the sale, and the Forest Service withdrew it. Later that same year, the Forest Service proposed a management plan for the area that called for multiple use, including timber harvest. That plan also raised opposition, and after a series of public meetings, the forest supervisor appointed a citizens' committee representing a range of local opinion. The committee endorsed the Forest Service plan, and it was adopted. Environmental groups appealed the plan twice, but ultimately it was upheld by Forest Service Chief Edward Cliff.

The chief's decision sparked protests, including a demonstration by a student group calling itself Nature's Conspiracy at the forest supervisor's office in Eugene.[45] A group of loggers showed up to offer their own opinions, and the situation was tense for a while. The student protesters wanted Forest Supervisor David Gibney to come out and talk to them. He didn't, but said later that he'd invited the demonstrators inside and they had refused to come.[46] (Rosboro's forester, Dave Burwell, evidently was there, for he later said he thought that if Supervisor Gibney had gone out to meet the demonstrators, there would have been a fight between the students and the loggers.[47]) The authors of a history of the French Pete controversy conclude that "the nucleus of the [protest] group was made up of students and faculty members mobilized by the Sierra Club and affiliated organizations, with a sizeable number of students simply coming along to see the show."[48]

The controversy over French Pete continued for another decade. After much more official study and public commentary, the area was added to the Three Sisters Wilderness in 1978.

The passage of the Wilderness Act and the French Pete tug-of-war exemplified the growing divide between old-fashioned "conservation," meaning wise human use and management of natural resources, and the more radical notion that human manipulation of nature is nearly always unwise—a philosophy that underpins much modern environmentalist thought. The divergence of these ideas in the mid-1960s took timber companies like Rosboro by surprise. The idea that timber was a crop, to be scientifically managed for human benefit—how could anyone argue with such a sensible proposition? And yet suddenly the whole world seemed to be doing just that.

Unlike Eugene, which was developing a reputation for radical activism, conservative Springfield managed to hang onto its small-town 1950s flavor, at least for a while. In 1960 Springfield was still proud to be known as Lunch Bucket City. "Lumber was still king, Main Street still had department stores (although sales were down), and giant malls were just a glint in some far-sighted developer's eye."[49] (Neighboring Eugene was calling itself "Lumber Capital of America," and Paul Cole remembers that some drivers sported license plate holders carrying the message.)

Springfield had its track star, Bill Dellinger, a Thurston High track coach who won the bronze medal for the 5,000-meter race at the Tokyo Olympics in 1964. It had its quirky local author, Ken Kesey, who was reaping international fame for his novels *One Flew Over the Cuckoo's Nest* and *Sometimes A Great Notion,*[50] the former a vivid elaboration on the theme of a sick and crazy system (entirely in keeping with the spirit of the times), the latter an elegy on the Paul Bunyanesque figure of the "gyppo" logger, by then largely gone from the Oregon woods.

Troubles in the world notwithstanding, the view from inside Rosboro Lumber Company was decidedly optimistic. The new stud mill was completed in 1975 and soon was sawing 94,000 board feet per day, 21.2 million board feet per year. The big dimension mill was sawing 241,000 board feet per day, 55.8 million board feet per year. The plywood plant was producing 174,000 square feet per day (⅜-inch basis), 41.6 million feet per year. Glulam was producing 77,500 board feet per day, 17.2 million board feet per year.

Wage and price controls had been lifted, inflation was abating, and prices for wood products were starting on an uphill climb, albeit a bumpy one.[51] "I think one of the most simple ways to establish a frame of optimism for the future of the industry is simply to clarify where the industry is, and then draw your own conclusions," declared Val Gardner in his June 7, 1968 *News Burro*, written two days after Robert F. Kennedy was assassinated. "A comparison would be that a young man 20 years old certainly and obviously has a greater future than a man who is 60 years old, and the forest products industry really is an infant industry.... An exciting and most interesting future is coming to us."[52] If he were alive today, Gardner might be surprised at how prophetic his words turned out to be.

Notes

1 Val Gardner, The Rosboro *News Burro*, Aug. 6, 1965.

2 Val Gardner, *News Burro*, March 15, 1968.

3 Val Gardner, *News Burro*, May 28, 1965.

4 Val Gardner, *News Burro*, June 11, 1965.

5 "This is Rosboro," (company pamphlet published about 1965), p. 16.

6 Gardner, *News Burro*, March 15, 1968, graph p. 2.

7 Adams, Thomas C. *Log Prices in Western Washington and Northwest Oregon, 1963-68* (Portland, OR: Research Note PNW-127, USDA Forest Service Pacific Northwest Forest and Range Experiment Station, 1970).

8 Interview with Paul Ehinger, Feb. 23, 2006.

9 *Random Lengths Yearbook*, 1966 edition, p. 93.

10 *Random Lengths Yearbook*, 1969 edition, p. 104.

11 *Random Lengths Yearbook*, 1973 edition, p. 168; 1978 edition, p. 184.

12 *Random Lengths Yearbook*, 1969 edition, p. 104.

13 Gardner, *News Burro*, Oct. 29, 1965.

14 Newspaper clipping from Rosboro scrapbook, paper unnamed, dated Nov. 1967.

15 Gardner, *News Burro*, June 13, 1966.

16 Interview with Ira Lowe, Aug. 10, 2005.

17 Interview with Clyde Hughes, Oct. 14, 2005.

18 Gardner, *News Burro*, Feb. 20, 1967.

19 Two decades later, Paul Cole adds, one of Rosboro's competitors would go all-out to build four new glulam plants, with the goal of adding capacity of 250 million board feet, at a time when the whole U.S. market absorbed only 300 million board feet.

20 Gardner, *News Burro*, Sept. 15, 1967.

21 Gardner, *News Burro*, Aug. 6, 1965.

22 Interview with Ira Lowe, Aug. 10, 2005.

23 Ibid., Lowe interview, Aug. 10, 2005.

24 Gardner, *News Burro*, Nov. 1, 1968.

25 Maxi Mill, http://www.maximill.com/. Consulted March 2, 2006.

26 Interviews with Paul Cole, July 8, 2005, and Ira Lowe, Aug. 10, 2005.

27 Ibid., Lowe interview, Aug. 10, 2005. Lowe said, "Not only Paul and me, but the woods people."

28 Ibid., Cole Chapter 5 comments, May 26, 2006.

29 "Rosboro Family Tree." Unpublished company document.

30 Interview with Gene Scherffius, Feb. 15, 2006.

31 "Rosboro to end its high-lead logging operations." Springfield *News*, April 1, 1968.

32 Ibid, Scherffius interview, Feb. 15, 2006.

33 Ibid., Scherffius interview, Feb. 15, 2006.

34 Gene Scherffius, comments on draft 3, April 21, 2007.

35 Ibid., Scherffius interview, Feb. 15, 2006.

36 Ibid., Scherffius, comments on draft 3, April 21, 2007.

37 Paul Cole memo, Aug. 3, 2005.

38 Interview with David Weza, Aug. 30, 2005.

39 Gardner, *News Burro*, Sept. 3, 1965.

40 Gardner, *News Burro*, July 21, 1967.

41 Gardner, *News Burro*, March 12, 1968. The measure passed, and construction on the new Oregon State forestry building began in 1970. It was named Peavy Hall after the school's first dean, George Peavy.

42 Paul Cole taped notes, up to Feb. 7, 2006.

43 Rakestraw, Lawrence, and Mary Rakestraw. *History of the Willamette National Forest* (Eugene: USDA Forest Service, Willamette National Forest, 1991), p. iv.

44 Ibid., Rakestraw and Rakestraw, p. 115.

45 Ibid., Rakestraw and Rakestraw, pp. 111-115.

46 Ibid., Rakestraw and Rakestraw, fn. 70, p. 121

47 Ibid., Rakestraw and Rakestraw, fn. 121.

48 Ibid., Rakestraw and Rakestraw, p. 115.

49 Velasco, Dorothy, and Mara Velasco. *Springfield: Between Two Rivers* (Montgomery, AL: Community Communications, Inc., 1999), p. 76.

50 Ibid., Velasco and Velasco, p. 79.

51 U.S. Department of Labor. "Softwood Plywood Price Index" (graph, p. 23). From Paul Ehinger, Feb. 23, 2006.

52 Gardner, *News Burro*, June 7, 1968.

6

TURBULENCE WITHIN AND WITHOUT

"With the type of company we are—and our timber base—we expect to be one of the last of the independents around."

—*Paul Cole*

WHEN VAL GARDNER retired in 1971, the *News Burro*, his periodic newsletter to the company's wholesale customers, retired with him. In 1977 the *Newsboro* name was revived (with a different spelling) as a newsletter aimed at company employees. The editor, Larry Kramer, promised to highlight the leisure-time and family activities of Rosboro employees, spreading the word about who shot the biggest buck and who caught the biggest fish. "The newest proud parents, or smiling newlywed will likely find himself blinded by our camera flash," he wrote.[1]

The newsletter's premier issue featured the first in a series of articles titled "You Call it Logging," about the company's woods operations. "If you ask a question about forestry or logging around here," Kramer began cheerily, "chances are someone will point you in the direction of Dave Burwell."

The series ran through the next four issues, highlighting Burwell's uphill-felling technique, introducing the woods crew, explaining the yarding machinery, and reporting on the Sugar Creek forest fire, which had recently burned a steep 30-acre stand of timber on Rosboro's Finn Rock tree farm. Kramer closed his stories with these words: "Please remember the danger when you use the forest the rest of this summer and fall."[2]

Rosboro Lumber Company was entering a comfortable middle age. Old-timers who'd been with the company since the early days were beginning to retire. Frank Diggs was one of the earliest; he had followed Whit Rosborough out from Arkansas in 1939. He'd been hired by Beuford Cole to help build the sawmill, and when that was done Cole put Diggs to work on the powerhouse crew. Diggs worked at Rosboro for almost thirty-eight years before he retired in 1977.[3] Diggs' son Larry joined the company about 1960 and was still there twenty years later, driving a lift truck in the plywood plant and serving as president of Local 2750 of the Lumber and Sawmill Workers' Union.

THE NEWSBORO
FEB-MAR 1977

Our First Issue

You hold the very first issue of THE NEWSBORO, a magazine for Rosboro people and about Rosboro people.

Because this magazine is something new, you've probably got some questions, so this is a good time to give some answers to a few of the most likely ones.

In the future you'll be receiving your copy with the mail every other month. Our first issue arrived in your home during February, while the next one will come to you in April.

Who gets a copy? Everyone on our payroll gets a copy. The mailing list comes right out of the records, and magazines are sent directly to the address listed in your file. So you don't have to subscribe or pay for it.

So, why have a Rosboro magazine you ask? Well, our biggest goal is to spread a bit of recognition around about the many of you who deserve some notice, especially about the large events and small that make up the good things of life. We want to say again, straight out, this is your magazine, designed for and about all of you in the many Rosboro operations. It is true that at times when issues challenging to our timber industry and to each of us come along, there will be comments made in these pages. The comments made here will be plainly put, in as few words as necessary.

Mostly though, we will be talking to, and writing about what active Rosboro people are doing. For example, some of you are working out in the community with civic, social or public groups, solving problems, or promoting local public events. We want to meet you.

We'll be bringing you word about people in hot pursuit of a private pilot's license, and some information about how it is done.

We expect whoever brings back the biggest buck or bull of the season to be full of news about how he did it. That goes too for fishermen of all persuasions who bring back trophy trout or shiny steelhead.

The newest proud parent, or smiling newlywed will likely find himself blinded by our camera flash, while he relates a detail or two.

Rosboro Lumber Company has a history as colorful as the Texan, T.W. Rosborough, who built a mill and established the town of Rosboro, Arkansas more than 70 years ago. How his first mill grew, and how our Rosboro operation first came west to Springfield in 1939 and 1940, is a story we'll tell in future issues. A prime part of that story will come from the recollections of workers, like Frank Diggs, and Bill Kocer, who have been here since those early beginnings.

In this issue we begin a study of the woods side of Rosboro, with forester Dave Burwell as our guide. From time to time THE NEWSBORO will feature the uses of our fine manufactured wood products. Also, we'll say a few words about safety, about the divisions with good records who deserve credit for their efforts.

Well, all this is probably too many words already about our plans. We just want to urge everyone around Rosboro to join in making this publication, which is really yours, fun and informative.

Why Name it THE NEWSBORO?

Once upon a time (more years ago than those who remember the details like to say) a fine, fact-filled fellow by the name of Val Gardener held down a sales desk in the Rosboro office. Val had a flair for words, and figured that his lumber customers probably would remember the name Rosboro better if they had something printed in their hands telling of the company. So he promptly worked up a regular little newsletter and sent it off to his customers. It contained tidbits of company news and facts about some products we produced. Unlike our new magazine, his never reached the plant crew.

He called his newsletter The Newboro. In our next issue we hope to get the whole story on that first paper from Val Gardener, a recent retiree, himself. If we can pry him loose from the golf course to reply to our letters.

Walt Burge Calls it a Career

Those three tall stacks sprouting from the boilers under the roof of our powerhouse pinpoint the heart of Rosboro's manufacturing operations in Springfield. The powerhouse, and the men who've watched over those mammoth steam and energy engines, first put the power to the machinery of our sawmill when it was built brand new over 37 years ago.

According to Earl Jenks, who should know since among his several jobs is one with the title Powerhouse Supervisor, it wasn't long ago our powerhouse crew could count something close to 200 years of experience combined among the several-man crew. With some recent retirements in the past couple of years, that total's down somewhat. When Walt Burge hit the retirement route on the last day of January this year, he took close to 34 years of experience along with him.

Not that Walt was any rank newcomer when he walked into our powerhouse for the first time that summer back in 1943. Hardly. "My first powerhouse job", said Walt, "was over in Wendling, with the big Booth Kelly mill." As he talked, just a day or two before retirement day, Walt recalled how it was for a Kansas bred midwesterner to

The *Newsboro* first issue, February-March 1977

Another notable employee nearing retirement was head sawyer Roy Edwards. Edwards and his wife had also followed the Rosborough operation out from Arkansas, setting forth on their wedding day to join her brother, Rosborough's woods foreman Vernon Williams, and his wife in Springfield. After a brief stint in the woods, Edwards came to work in the mill. He became pony-rig sawyer in 1950 and head sawyer in 1960.[4] "In those days the head sawyer had to make all the decisions [about how to break down a log] immediately," said Edwards, 88, who retired in 1984 after forty-four years with the company. "For years I could figure faster than a calculator. Now computers do all that." Edwards remembers when computers were first installed in the dimension mill: "The first computer took up the whole wall next to the carriage."[5]

Rosboro's long seniority list was a tribute to the financial health and good management of the company, in the view of Ira Lowe, then manager of lumber and veneer operations. "Here's a fellow who's been with us since 1947 … another since 1940 … 1946 … 1952 …" he noted in a 1980 *Newsboro* story. "There are not very many companies that have that many employees with that kind of longevity. … They are obviously skilled workers who could have worked any place they wanted to." Rosboro, he asserted, had "as stable a crew as any other plant. That's what a company is … it's people. It isn't machines and it isn't buildings. A company is made up of people and the ones with the good people—the responsible ones that are willing to work as a team—are the ones that will survive."

New Finn Rock bridge, built in the 1960s

As the old-timers departed, their places were filled with younger men and, increasingly, women. A 1980 issue of *Newsboro* has photos of half-a-dozen women pulling veneer and working in the plywood plant, as well as performing more-conventional office jobs.[6] Louise McQuin, who had been with Rosboro since 1942, had worked her way up from general office duties and now sold plywood for the company. In 1980 McQuin was named Lumberwoman of the Year by the National Convention of Hoo Hoo-Ettes, the women's auxiliary to the Hoo Hoo Club, a wood-products industry organization.

McQuin's position was unusual in the industry, but she couldn't see why selling plywood had to be man's job. "Years ago," she said in a 1980 interview, "they thought you had to work in a mill before you could sell. Now they hire lots of salespersons who have never seen the inside of a mill. You don't even have to know how to pound nails."[7]

Outreach to Employees

The resurrection of *The Newsboro* in the spring of 1977 may have been, among other objectives, an attempt by the company to reach out to employees with some preemptive communication. Labor-management relations were tense at times during those years. There'd been an ugly incident in the veneer plant on Oct. 26, 1976. Someone—it was never revealed who—placed a handful of bolts inside a gearbox in the veneer-pulling area, apparently with the intent of jamming the machinery and forcing a shutdown. The company declared it an act of sabotage.[8] The incident led to the firing of 10 employees, a union lawsuit against the company, and a period of troubled relationships between Rosboro and Local 2750 of the Lumber and Sawmill Workers' Union, which had represented Rosboro's employees since the early days of the company.

In the company's view, the sabotage was the final straw in more than a year's worth of poor performance by some green-end crew members who, as Paul Cole put it, "didn't want to work." Veneer-pulling chains were getting broken, the line was stopping too many times for no good reason, good veneer was being destroyed by careless handling. There was too much horseplay, and, in the company's view, a generally poor attitude prevailed.

As soon as the incident was reported, Paul Cole ordered the veneer plant shut down and sent twenty-eight of Rosboro's approximately three hundred employees home.[9] Early in November he and Rosboro's labor consultant, Jack Gossard, met with union leaders.[10]

A week later Cole wrote a letter to Rosboro employees. His tone was somber but firm. "It is with the deepest regret that I sit down today to write this letter," he began. In the past weeks "a department of the company operations has been plagued with slow downs and petty work stoppages, culminating in an act which is difficult to look upon as anything other than deliberate sabotage. This is a lawless act and as we are all aware, no society, institution or even family unit can function when anarchy is at work within it."

Cole's letter entreated the union leadership to "rid the union of those elements which are creating an atmosphere of lawlessness which in the long run no one can tolerate and to work with us in an effort to solve this problem and return the plant to normal operations." If cooperation were not forthcoming, he said, the company would consider its options, which "are varied, but all involve a great deal of hardship to the employees in all phases of our operations."[11]

Cole had already made the company's position plain in an Oct. 29 memo that was circulated to union leaders. Unless the green-end employees turned the culprits in, the company would be within its rights to shut down the green-end plant and even the plywood plant if it chose.

On Nov. 13 the union held a special meeting to show the company that it did not condone acts of sabotage. Union members voted to put up a reward of $1,000 for anyone who identified the saboteur, but no one came forward. A week later the company's stockpile of veneer was exhausted. Rosboro closed the plywood plant and sent another eighty-nine employees home.[12]

This was an overreaction, in the view of union official Jim Bledsoe, executive secretary of the Lumber and Sawmill Workers' parent organization, the Western Council of Lumber Production and Industrial Workers. "I think sabotage is over-dramatizing what happened at that plant," he said. He blamed company supervisors for any crew misbehavior[13] and maintained that the union should not have to "act as a police force."[14]

A second letter from Cole to employees on Nov. 22 noted that no one had yet come forward with information about the alleged saboteurs. His queries of other employees and their families, he wrote, had revealed that "the situation is much worse than I first thought." He thanked those employees and family members who had expressed concern and support for the company, but warned that the company was considering continuing the curtailment, which would result in the layoff of fifty to ninety employees—although "[W]e are reluctant to do this because the innocent will suffer along with the guilty." Another option, he said, was to fire the crew members reputed to be the troublemakers.[15]

On Dec. 9, the company fired ten men and the plywood superintendent. Four days later it reopened the veneer plant and called the twenty-eight employees back to work.[16] Production immediately rose 40 percent, according to company figures.[17]

The union sued the company, alleging that the plant closure was an illegal lockout and that the employees had been fired without just cause.[18] The union wanted the company to open the plant again, reinstate the fired employees, and give them back pay. Rosboro maintained that the shutdown was not an illegal lockout—that the company had an obligation to protect the safety of its employees by shutting down damaged machinery.[19]

In January the union took a vote on whether to strike over the firing of two more employees. The strike vote failed, but there were clearly hard feelings on both sides.

Over the next few months, Rosboro negotiated settlements with seven of the discharged employees. U.S. District Judge Otto Skopil eventually heard the cases of Clinton Harris and Ed Kubik, the two employees who had not settled with the company. The judge awarded $1,614 in back pay to Kubik, who he said was "a concerned, conscientious employee" who should not have been fired. The judge denied benefits to Harris, saying that his testimony was not credible because Harris had falsely claimed he had been unable to find other work since being fired.[20]

The sabotage incident left hurt feelings all around, and it soured relations between the company and the union. In fact, the plywood employees[21] did go on strike in August of 1978, Rosboro's first since the massive industry-wide strike of 1954. Employees of the other plants joined the walkout.[22] This strike was not related to the sabotage incident, nor did it have to do with wages and benefits. Rather, the issue was new equipment that automated manufacturing processes and forced changes in employees' job requirements.[23]

Rosboro's program of equipment upgrading and modernization had gotten into full swing in the late 1960s. The resulting improvements in efficiency led inevitably to leaner crews. However, said Ira Lowe, the company tried not to let people go before they were ready. "We have always tried in the past modernization not to lay people off but to let the natural attrition of people leaving, retiring and so forth, take place and move these people that have been displaced into those jobs."[24]

Yet there were still sticking points. After hard negotiations, the company and the union agreed to seek help from the Federal Mediation and Conciliation Service. They reached an agreement that, according to Paul Cole, the union later breached.[25] "This became an important issue," he said, recalling the negotiations at that time. "We maintained that an agreement had been reached, but the union proceeded to push for a strike anyway. In our dealings with employees and the news media, we kept repeating, 'The union reached agreement with us; there should be no strike.'"

The company sent a letter to employees' wives at the beginning of the strike, asking them to persuade their husbands to stop striking and go back to work. The strike ended within a month, and the letter may have hastened the settlement.[26] It was sent over Paul Cole's signature, but it was actually penned by Nonnie Cole, who has, said her husband, a remarkable way with words. When someone in the union later complimented Rosboro's labor consultant, Jack Gossard, on his skill in writing such an effective letter, Gossard demurred, "You're giving me credit I don't deserve."[27]

Paul Cole, Gordon Culbertson, Mike McCrady, Ira Lowe, and David Weza in the early 1980s

Cole believes the problem was created by the union leadership at the time. "We felt we had bargained openly and fairly, reached an agreement, and then witnessed the union leadership lead their members into a strike that should not have happened."[28] The labor difficulties led Cole to redouble the company's efforts to smooth relationships with its employees. It was at this time that Rosboro, at Jack Gossard's urging, created the *Newsboro*. Almost every other lumber company had such a newsletter, and Rosboro didn't.[29] An employee newsletter was an important tool: it helped to convey company information to employees through interviews with managers and supervisors and through regular columns such as Cole's "As I See It."

Also during that time, Rosboro's safety program was pushed to a higher profile. The company hired consultant Joe Dolezal to train employees in good safety practices. The first company-sponsored safety banquet, held in the spring of 1979, rewarded 200 employees in the company's safest divisions with awards, a gala dinner, and door prizes such as a power bench grinder and a compact TV set.[30]

Rosboro continued to work toward a team-style management system. The overall company team was composed of the division managers: Ira Lowe (lumber and green veneer), Clyde Hughes (glulam) Gene Scherffius (timberlands), Don Cooley (plywood), and general manager Paul Cole. Each division had its own management team, composed of the manager and all his supervisors.[31]

The company hired consultants Jack Gossard and Chuck Pyron to train managers and supervisors in the concepts and practices of teamwork. The Gossard-Pyron workshops, which were well known and highly regarded in the Eugene business community,[32] covered motivation, communication, leadership, decision-making, listening, helping, and "giving performance feedback and setting improvement goals."[33] A 1981 company-produced training film shows a fresh-faced David Weza, recently promoted to assistant operations manager, earnestly explaining company policy to the employees. Weza later became Rosboro's chief executive officer, but he still has a youthful appearance that belies his fifty-two years.

In a 1979 *Newsboro* article, Paul Cole explained his vision for team-style management:

> ...active communication ...by talking and listening face to face. As the team talks and listens, everyone helps to consider and discuss ways to improve the relationships and understanding between themselves and employees. ... Everyone in the team meeting participates. ...This is true communication, because everyone who talks is honestly being listened to. In this way of open communication the team reaches a consensus. In other words, an understanding and an agreement is reached harmoniously by shared talking and listening.[34]

With Rosboro, as with most businesses at the time, team-style management was easier to preach than it was to practice. Cole and his managers were of a certain generation and a certain temperament, and they were most comfortable with a top-down, chain-of-command style of running things. David Weza was a supervisor in Mill B at the time the team-management effort got rolling. He speculates that Paul Cole didn't really believe, deep down, that team-based decision-making would work. "He believed that, when push comes to shove, somebody has got to make a decision. So, to call it a team approach at that time really isn't accurate. Everybody had a role, but the manager was in charge."[35] A decade later the company would renew its commitment to team management. Eventually, by most accounts, Rosboro's team-management processes became functional and highly effective.[36]

Edger for small-log mill at Feather River

Overseas Competition

The turbulence within the company during the mid- to late 1970s was mirrored by turbulence without. Events of the previous decade had set the stage for major dislocations to the wood-products economy. Now the road was starting to get seriously bumpy.

One looming threat was the increasingly heavy competition for logs from Japan, which was continuing to building itself up after the devastation of World War II. Japanese builders and consumers preferred to import raw material and mill the lumber to their own specifications rather than buy finished products from the United States.

Exports of raw logs from Oregon to Japan began in 1961 with a volume of about 120 million board feet.[37] By 1979 the Japanese market was soaking up 618,000 million board feet of Oregon logs. Log exports would become a divisive political issue in the late 1980s and early '90s. In the early '70s, however, the export trade was perceived merely as a lucrative market for those companies lucky enough to have logs to sell, for there was more profit to be made in exporting them to Japan than in sawing them into boards at home.

Stacks of veneer

Pacific Northwest mills were also struggling for domestic market share. Canadian lumber was emerging as a competitive threat, and so was Southern pine lumber, as the forests of the South, extensively logged between 1900 and 1950, began to produce marketable timber again.

Another burgeoning development was the designation of more and more federal forest lands for primarily recreational purposes—something the timber industry generally viewed with alarm. An article in *The Newsboro* for August-September of 1977 mentions that the Forest Service was evaluating more than 150 large timbered areas in Oregon for addition to the wilderness system. "Of the 14 million acres already preserved in Wilderness, more than 10 million are in commercial timber! We believe in Wilderness but the question is How much is enough?" Employees were exhorted to write their opinions on an enclosed sheet of paper and give it to their supervisors for inclusion in a packet of comments to be sent to the Forest Service.[38]

In another *Newsboro* issue, company forester Dave Burwell spun a vivid scenario of a hypothetical wildfire that, in his view, was likely to strike the forests of the Cascades and Coast Range if they were "left to their natural life progression" rather than being roaded and logged. What if hikers in the newly designated Middle Santiam wilderness failed to drench their campfire on a hot summer day? "A great disaster" would ensue, predicted Burwell—a holocaust that would "push ... the Tillamook burn into insignificance as forests, mills and homes, all of Oregon's prosperity, went up in black smoke."

Burwell concluded with a summary of his commonsense conservationist's creed, which so many people seemed to be questioning these days: "Either man manages the forest to supply useful needs, or nature will do the job in her historic way. You can no more preserve a forest forever than you can a field of wheat. The harvest will come in a rational, thought-out manner, or it will come through the destruction of fire."[39]

On top of all this, the whole country was struggling with a stagnating economy coupled with crippling inflation. In 1973, oil-producing Arab countries imposed an embargo on shipments of crude oil to the West. Arab leaders resented U.S. support for Israel in its conflict with Egypt and Syria in the Yom Kippur War, fought in October of that year, and the embargo was their punishment.

Suddenly the price of crude oil doubled, causing massive fuel shortages throughout the United States. At the height of the crisis, cars could be seen lined up all the way around the block at gas stations across the country. For a time gasoline was available only on alternate days (for example, odd-numbered days for odd-numbered license plates, and even-numbered days for even-numbered license plates), and sometimes not at all. The oil crisis aggravated a worldwide economic downturn already under way and prompted U.S. economists to coin a new word: "stagflation"—a vivid if lumpy catchword for the country's economic situation in the mid-1970s.[40]

The wood-products economy saw a downturn in 1970 and another in 1974-75,[41] the latter associated with the oil embargo. The fuel shortage made life especially difficult for West Coast mills because their distance from major Eastern markets made them pay higher shipping costs.[42]

By the end of the decade Oregon lumber production had rebounded to pre-recession levels. In 1978, plywood production reached a peak and also enjoyed its last hurrah: the 1.9 billion square feet of plywood produced in 1978 amounted to the fourth-highest production in any year since 1945, and production has been declining ever since.[43]

The uptick at the end of the 1970s was short-lived. The whole world had changed since the good times of the mid-'60s. The Japanese were buying good logs and paying top dollar for them. Cheap lumber from Canada and the South was flowing into the domestic marketplace. The vast federal forests that timber companies had always assumed were theirs—and on which they had become vitally dependent—began to be chipped away as tracts of forest were designated as wilderness or otherwise withdrawn from logging. Ruinous interest rates discouraged economic expansion, including home construction, historically the bread and butter of the wood-products trade.

In October 1979, the bottom fell out of the wood-products market, and over the next three years lumber prices dropped more than 48 percent. The anxiety of the time was captured in an article from *Random Lengths*, a lumber-industry trade journal: "Caught in a maelstrom of soaring interest rates, slumping housing starts, plunging lumber and plywood prices, production cutbacks, a bitter election battle, and even an exploding volcano,[44] the industry somehow managed to stagger through 1980."[45]

The recession of the early 1980s was a nationwide phenomenon, but it hit timber country particularly hard. A wave of mill closures throughout the Pacific Northwest left many workers jobless. In some towns the unemployment rate topped 25 percent; in Springfield, 10 percent of workers were unemployed—bad enough. The effects of the shutdowns rippled through communities, bringing hard times to retailers and other businesses, slashing local tax revenues, and starving community-service organizations.

The smaller, family-owned mills were the most vulnerable to hard times. Typically undercapitalized compared to the giants like Weyerhaeuser and Georgia-Pacific, many of them didn't own timberlands and didn't have the resources to bid against the big companies for government timber. In a 1977 article in *The Newsboro*, Paul Cole named six lumber mills in western Oregon that had closed their doors: Hills Creek Lumber, Mount June Lumber, Star Lumber, Litchy-Gregory Lumber, Barker Willamette, and Hines Lumber. "In most cases," he wrote, "their equipment was sold off piecemeal. In a couple of instances other companies took over a tiny fraction of their productive capacity."[46] The American Plywood Association reported 17 mill closures and 18 curtailments in Oregon in mid-1980.[47]

Big companies coped by retooling, merging, or migrating. Georgia-Pacific Corp., which had come to Portland from the South in 1954, moved its headquarters back to Georgia in 1982, and the green-neon "G-P" logo atop the company's landmark Portland office building winked out for (probably) the last time.[48]

Outside Mill A

The shakeout triggered by the '80s recession lasted four painful years. It may have been inevitable. Timber's peak production years were over; companies had developed more manufacturing capacity than they could sustain, and after the heady 1950s the industry settled into a gradual decline that steepened through the 1970s.

In addition, many Northwest mills, built in the 1950s or earlier, were reaching the end of their productive life. To stay competitive, companies had to invest in expensive equipment upgrades and new technology. An economic crisis is not an easy time to spend money on capital improvements, but the companies that could afford to make these investments did so, improving their odds of survival—and also boosting productivity to the point where far fewer employees were needed. By the time the national economy recovered in the mid-1980s, 48,000 timber-related jobs in the Pacific Northwest were gone, most of them never to return.

In the late spring of 1980, Rosboro was running at about 65 percent capacity in its lumber mills, and Paul Cole expected that figure to drop to 50 percent. Mill A, the dimension sawmill, had just undergone a major overhaul and was running at below capacity. The plywood plant was operating less than four days a week.[49] Rosboro was juggling its several product lines to keep the doors open as best it could. "We have five different kinds of businesses," said Cole. "We have lumber, veneer, plywood, glulam, and timber, and they have somewhat different business cycles for each."[50] That diversity plus the company's timberlands, he said, put Rosboro in the best possible competitive position during these hard times. "With the type of company we are—and our timber base—we expect to be one of the last of the independents around."[51]

Timber Crisis

The Pacific Northwest lumber companies that survived the shakeout learned one very important thing: Whatever you do, be sure you have enough timber. Between 1977 and 1980, Rosboro and many other companies had bought federal timber at the prevailing exorbitant prices—as much as $500 a thousand board feet[52]—to assure themselves of a supply.

Then the economy went sour, and companies knew they wouldn't be able to sell the lumber for enough to cover the cost of the logs. "Rosboro had 200 million board feet of that high-priced timber under contract," said Gordon Culbertson, then Rosboro's log and timber buyer. "We were sawing and peeling about 75 to 80 million board feet of logs a year. We'd bought the timber at $450 to $500 per thousand board feet, and now [1982] the price was around $120 per thousand."

Industry leaders in the Northwest, with the backing of Oregon Senator Mark Hatfield, mounted a legislative campaign to seek relief from what they saw as an unbearable burden. In July of 1983, President Ronald Reagan authorized the Forest Service to grant five-year, no-interest extensions on contract deadlines to keep companies from defaulting. These extensions offered some relief, but they were not a full solution—no one foresaw a day when those high-priced logs could be profitably milled.

The following year Congress passed a law that allowed the federal government to terminate some of the contracts. As he signed the bill in October of 1984, President Reagan stated, "It now appears that ... market prices for wood products will not rise enough to prevent severe economic hardship to many forest-dependent communities unless the provisions of this new law are implemented. ... Under all the circumstances, the legislation provides a balanced approach to resolving a many-faceted and difficult problem."[53] The law permitted companies holding high-priced contracts to "turn back," or withdraw from, a substantial fraction of them for a few cents on the dollar and extend the deadlines for the remainder. The amount a company could turn back depended on how much it had under contract, when the contracts were negotiated, and the size of the company. At the time some 20 billion board feet of federal timber was under contract, and in the end about 8 billion was turned back.[54]

The so-called "timber buyout" was controversial. Environmental groups assailed it as a bailout of the timber industry—"and that's the only term you can use for it because they got out of $3 billion worth of timber contracts for $200 million," said Robert Wolf, a forester and analyst for the Congressional Research Service in a 1989 interview.[55,56] Wolf, who conducted some of the financial analysis of the 1984 law, maintained that the companies holding the high-priced contracts had overbid themselves trying to bury their competitors, confident that the economy would continue to expand, or, if it didn't, that the government would relieve them of their contract obligations. "The industry was bidding, secure in the knowledge that if things got rough, they could get extensions [of contract deadlines] and they cost them nothing. You had a small down payment, you held the timber, and you got an extension. ... So they didn't believe they were going to be hurt by the high bids."[57]

The timber buyout law raised public awareness about the issue of unprofitable federal timber sales. Research by Wolf and others revealed that quite a few national forests were losing money on their timber.[58] Congress, led by Rep. Sid Yates (D.-Illinois), began to probe the accounting practices of the Forest Service. In 1983 a subcommittee chaired by Yates, who was chairman of the House Appropriations Subcommittee on the Department of the Interior and Related Agencies, requested data from each national forest on timber sale costs and revenues. Subsequent analysis revealed that revenues from many National Forest timber sales did not cover the cost of preparing them.[59] These findings led Congress to demand, in its later appropriation bills, that the Forest Service develop a more accurate accounting system.

Courtesy of Gordon Culbertson

Logging the Hull lands, Section 8

By the time the contract-relief bill was signed, Rosboro was embarked on an aggressive program to acquire timber from private lands and to buy more timberlands of its own. It was about this time that Gordon Culbertson came on board as a log and timber buyer. Culbertson was a young timber manager with Tangfeldt Wood Products, a customer and competitor of Rosboro's. He arrived at Rosboro in 1983, the year before the timber-relief law was enacted, when the company was seriously worried about the mortgages on its future. "If relief wasn't granted to mill operators by both the BLM and the Forest Service," stated Paul Cole flatly in *The Newsboro,* "many companies would go broke and cease operations."[60]

Culbertson does not agree that the timber-sale relief bill was an unfair subsidy to the industry. "To people who have never purchased timber, it may seem like a subsidy," he said. "And other parts of the country felt, well, it's the Northwest's problem. But it was a very tenuous time for the industry—very hard. The Northwest industry has always been tough and competitive," but this time the burden was too heavy. The buyout was necessary, Culbertson maintained, to ease a vital economic sector through a particularly rough time.

Within a year of his hiring Culbertson was promoted to timber manager, with responsibility for all the company's logging and timber operations. By then several of the federal timber sales Rosboro had been allowed to release came onto the market again, and Culbertson bought some of them back at much-reduced prices. He also began acquiring private lands for the company. One of Rosboro's first major land purchases was the Hull lands in the Coast Range near Bellfountain. Culbertson negotiated a $40 million handshake deal with Swanson-Superior Forest Products, based in Noti, on the eastern slope of Lane County's Coast Range.

The bargain was a good illustration of the competitive cooperation that had long existed among Oregon's family-owned mills. Like Rosboro, Swanson-Superior operated a small-log mill, but the company could not accommodate big logs as well as Rosboro could. "They approached us with the idea," said Culbertson. "They knew how well managed and financed we were, and they knew we'd be a good partner. It was a good deal for both of us." The two companies shared the purchase, dividing things up so that Rosboro got the smaller portion of the Hull land but took the bigger logs, and Swanson-Superior got the bigger portion of land but took the smaller logs.

Under Culbertson's direction, Rosboro went on to acquire timberlands from Weyerhaeuser, Pope & Talbot, Publishers Paper, and other companies with holdings in western Oregon. By the late 1990s the company owned about 75,000 acres of forest land.

South of the Border

By 1985 the national economy had begun to pick up a little. In that year two shuttered sawmills in Yuba County, California, came to the company's attention. The mills, built in the early 1950s, were geared for cutting old-growth white fir and pine. They had belonged to Diamond International Corp. until late in 1982, when the British industrialist and corporate raider Sir James Goldsmith acquired Diamond and dismantled the company, selling off many of its assets.[61] The two Yuba County mills were shut down. Obsolete and abandoned, they came onto the market at a bargain price.[62]

Cole, Gordon Culbertson, and Ira Lowe traveled to California to have a look at the mills, one in Oroville and the other in Linda, a small suburb of

Inside the Feather River mill

Marysville. They found rundown equipment and a sparse labor force—most workers had moved on after the mills closed. The Linda mill had been built in 1950 and not much upgraded since then. It was the better of the two, Lowe concluded, but even so, it would be a major rehab job. There'd be considerable rebuilding, refurbishing, and retraining to do. But these things could be done—Rosboro had been upgrading equipment and retraining employees in Springfield more or less constantly for two decades now, ever since Lowe had arrived.

The sweetest part of the deal, in Culbertson's view, was the abundance of white fir on the nearby Tahoe and Eldorado National Forests. These forests, first logged during the Gold Rush years, contained large stands of mature second growth, and it was cheaper than comparable raw material closer to home.

In addition, under its previous owner the Linda mill had been producing wide-dimension lumber comparable to what Rosboro was cutting at Springfield. "We felt, after looking at [the mill], that it had important things we needed, such as kiln-drying capacity," Culbertson said. "We felt we could adapt the mill based on our experience in Springfield."

After considerable discussion, Rosboro's managers agreed that the Linda mill looked like a good deal. They chose not to make an offer on the Oroville mill, a decision that would come back to haunt them.

Cole knew the company couldn't take such a major step without the blessing of the other owners. He and the management team made a special trip to Atlanta and laid the proposal before them.

Three more of the original owners had died: Tom Watts, of complications from Parkinson's disease in 1974, and Anna Watts Rosborough, widow of Whit Rosborough, in November of 1984, at the gallant age of ninety-four.[63] She was much loved, and her many friends and relatives continue to refer to her as "Miss Anna." Most tragically, Martha Watts had been murdered in her Atlanta apartment in December of 1978, by an intruder who apparently was after her jewelry.

The owners' executive committee was skeptical. Ruth Kainen and Jim Watts, in particular, were downright dubious.[64] It was a new thing for the company to extend itself so far out of its home territory—it was a bold step, maybe a reckless step. But these were uncertain times, Paul Cole knew. Things were moving fast, and the future was unpredictable. For a company in Rosboro's circumstances, playing it safe might be the riskiest course.

Ultimately the owners gave their blessing, and Rosboro Lumber Company acquired a subsidiary which they named Feather River Forest Products, consisting of one aging sawmill 430 miles from home.

In the summer of 1985 Culbertson and Lowe began to fly down to Marysville on alternating weeks: Culbertson one week, Lowe the next. Culbertson explored the timber situation, scouting what was available from both the Forest Service and private owners.

Lowe's task was to convert an old large-log pine mill to a small-log mill for white fir, hemlock, and Douglas fir. "Diamond had a system in [the mill] that worked only with logs of a certain size, plus or minus a little," said Lowe. "We needed to work with logs of all sizes." He rebuilt one side of the mill to handle logs of different species, 12 inches in diameter and under. "Ira wanted a machine that would saw nominal 2-inch pieces, and something that did not cost a lot," said Paul Cole.

Lowe decided that what was needed for this small-log side was a headrig consisting of a bank of circular saws—different from the "quad" bandsaw headrig in the small-log mill at Springfield. He bought a used double-arbor 12-inch gang edger, and this became the headrig for the small-log side. He devised a method to position the log, and also to hold the log as it was being sawed. He then upgraded the other side of the mill so that it handled the larger-diameter logs more efficiently. A new headrig carriage feed system was installed, and the edger and sorter were automated.

That winter the Yuba River rose in a flood and inundated the mill, which was still undergoing its painstaking restoration. "I'd just gotten home from Marysville," said Culbertson, "when Ira called me on the phone. He said 'Turn on your TV.' I hadn't been home an hour."

Culbertson tuned to a San Francisco cable channel and saw the levee breaking in Linda, California, "and there was our mill on TV." The flood swept through the mill and the office, leaving a foot-high stain on the wall. "It was a terrible mess," said Culbertson. "And all our work in the mill was lost."

Culbertson and Lowe doggedly went back to work. Soon they were joined by David Weza, who by that time was assistant operations manager of all the Springfield plants. Weza hired and trained plant workers and supervisors for the Linda mill, then went looking for a plant manager. The first two he hired didn't work out ("but they were colorful characters," he said). Then he hired Tom Karshneski, who succeeded in getting the mill to run consistently. Needing a manager and someone with a maintenance and engineering background, Karshneski hired Larry Works, who performed small miracles to keep the mill machinery going.[65] Don Bowker, who helped Mike Morrissey in lumber sales at Springfield, began to sell the products of the Feather River mill.[66]

Things were beginning to look up at Feather River. Along with effective managers, "we had a pretty good group of hourly employees," Culbertson said. "We started to get where we needed to be; we were heading where we'd hoped to be heading.

"But then came 1990."

From the Roseburg *News-Review*, May 19, 1991

The Owl Wars

By the mid-'80s, environmental and conservation groups had become vociferous opponents of logging on old-growth federal forests. They took their fight into federal court, suing the Forest Service and the BLM for failing to follow the law in protecting the environment. They won injunctions halting hundreds of timber sales, which would have produced a billion or more board feet of timber.[67]

Rosboro had its first brush with environmental protest in 1987. The issue was a 43-acre BLM timber sale called Bunker Hill, in the Mohawk Valley northeast of Springfield,[68] a tract of low-elevation old-growth Douglas fir. The Oregon Natural Resources Council and a local coalition, the Upper Mohawk Community Council, together organized a protest of the timber sale. A hand-lettered poster headed "Bunker Hill Stop The Destruction!"

advertised a gathering at Rosboro's Springfield office on January 2, the day before logging was initially scheduled.

The Upper Mohawk Community Council had earlier tried to persuade Lane County commissioners to buy or lease the 110 acres of forest that encompassed Bunker Hill and make it a public park. The commissioners said they couldn't afford to buy it, but would consider leasing it from the BLM if Rosboro defaulted on the Bunker Hill contract.[69] In his hand-written notes from a conversation with *Register-Guard* reporter Ed Kenyon, Paul Cole made clear his distaste for that idea: "BLM owns timber not us—we have to log. He asked loss if we defaulted—ans. hard to calculate. Not correct way to go for making a business decision. We felt co.'s should do all possible to not default." Testily, he scrawled, "Conflicting direction from County → log. Co. needs money so don't log put in park!"[70]

Courtesy of Gordon Culbertson

A northern spotted owl at Cane Creek on Finn Rock tree farm

The BLM turned down the county's application. The environmental groups appealed the decision and set a meeting for January 18. They asked Rosboro to delay cutting for 30 days.[71]

No, said Gordon Culbertson in a phone conversation with a member of the Upper Mohawk Community Council. "I told him that we couldn't alter our plans to fit their wishes. I explained to him that we had no choice but to honor the timber contract. … Assured him that I was attempting to be candid and straightforward and wouldn't expect negative public reaction."[72] Then Rosboro braced itself for a media event,[73] but the logging went forward with no protest.[74]

In June 1990 the U.S. Fish and Wildlife Service handed the environmental movement a significant victory when it placed the northern spotted owl (*Strix occidentalis caurina*) on the threatened list under the Endangered Species Act. The listing meant the owl's habitat of old-growth forests had to be protected from logging. And that, for timber companies, meant that the many millions of board feet of National Forest timber they'd been counting on were suddenly out of reach.

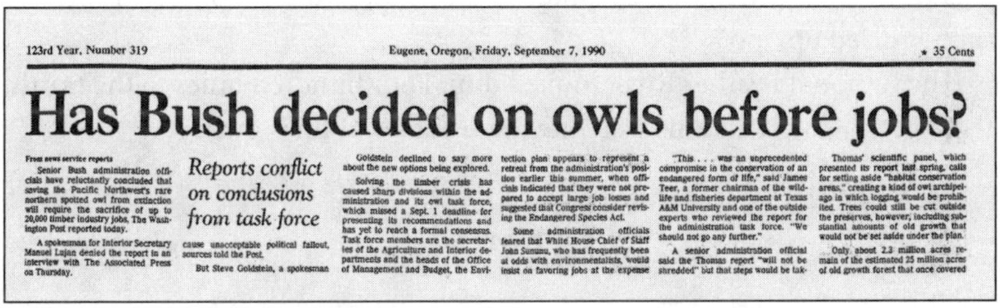

123rd Year, Number 319 — Eugene, Oregon, Friday, September 7, 1990 — ★ 35 Cents

Has Bush decided on owls before jobs?

Reports conflict on conclusions from task force

From news service reports

Senior Bush administration officials have reluctantly concluded that saving the Pacific Northwest's rare northern spotted owl from extinction will require the sacrifice of up to 20,000 timber industry jobs, The Washington Post reported today.

A spokesman for Interior Secretary Manuel Lujan denied the report in an interview with The Associated Press on Thursday.

cause unacceptable political fallout, sources told the Post.

But Steve Goldstein, a spokesman

Goldstein declined to say more about the new options being explored.

Solving the timber crisis has caused sharp divisions within the administration and its owl task force, which missed a Sept. 1 deadline for presenting its recommendations and has yet to reach a formal consensus. Task force members are the secretaries of the Agriculture and Interior departments and the heads of the Office of Management and Budget, the Envi-

tection plan appears to represent a retreat from the administration's position earlier this summer, when officials indicated that they were not prepared to accept large job losses and suggested that Congress consider revising the Endangered Species Act.

Some administration officials feared that White House Chief of Staff John Sununu, who has frequently been at odds with environmentalists, would insist on favoring jobs at the expense

"This … was an unprecedented compromise in the conservation of an endangered form of life," said James Teer, a former chairman of the wildlife and fisheries department at Texas A&M University and one of the outside experts who reviewed the report for the administration task force. "We should not go any further."

A senior administration official said the Thomas report "will not be shredded" but that steps would be tak-

Thomas' scientific panel, which presented its report last spring, calls for setting aside "habitat conservation areas," creating a kind of owl archipelago in which logging would be prohibited. Trees could still be cut outside the preserves, however, including substantial amounts of old growth that would not be set aside under the plan.

Only about 2.3 million acres remain of the estimated 25 million acres of old growth forest that once covered

From the Eugene *Register-Guard*, September 7, 1990

The listing thrust a small, shy denizen of old-growth Northwest forests into the national spotlight. Based on research by Forest Service and other scientists, the northern spotted owl, which inhabits older coniferous forests from the western Cascades to the ocean, was designated as an "indicator species"—an ecological thermometer used to gauge the well-being of the old-growth forests managed by the Forest Service and the BLM.[75]

The owl's listing further polarized the region into pro-environment and pro-timber factions. Loggers were mocked and insulted in some environmentalist newsletters, and bumper stickers were seen on pickup trucks that read, "I love spotted owls…fried."

To the fiercer spokesmen on the timber side of the political divide, the spotted owl stood for environmentalist duplicity and agency cowardice. "I am being put out of business by a surrogate species that will never be extinct and the goddamn preservationists know it just as well as I do," said a spokesman for mills in Oakridge and Creswell. "The spotted owl issue is absurd—it's a damn ploy to stop cutting trees."[76]

Former Rosboro timber manager Gordon Culbertson puts it more evenly: "The spotted owl was a symbol. The industry went from having plenty of timber to having almost none." Unlike several of its competitors, Rosboro, thanks to its large forest holdings, had more than enough timber to feed its Springfield operations.

Soon after the spotted owl's listing, a federal judge stopped timber sales on certain national forests and BLM lands. The ruling abruptly changed the picture in northern California, too, and suddenly the Feather River mill did not look like such a good deal. Instead of good sound white fir timber, Culbertson found himself bidding on salvage logs—decaying and full of insects. "That timber was not suitable for what we needed," he said, with some understatement. "At the same time, many companies that we'd been buying logs from were purchased by other companies, and our sources of private timber dried up, too."

The future of Feather River looked dim. Four hundred miles to the north, the Springfield operations were having troubles of their own.

Notes

1 The *Newsboro*, Feb.-Mar. 1977.

2 The *Newsboro*, Feb.-Mar. 1977, Apr.-May 1977, June-July 1977, and Aug.-Sept. 1977.

3 The *Newsboro*, Oct.-Nov. 1977.

4 Interview with Roy Edwards, Nov. 16, 2006.

5 The*Newsboro*, Apr.-May 1982; interview with Roy Edwards, Nov. 16, 2006.

6 The *Newsboro*, Mar.-Apr. 1980.

7 The *Newsboro*, July-Aug. 1980.

8 Paul Cole letter Nov. 4, 1976, attached schedule of events.

9 Lindsay, Lois. "Employe says Rosboro threat won't work" (Springfield *News*, Nov. 11, 1976).

10 In the United States District Court for the District of Oregon, Local 2750 Lumber and Sawmill Workers' Union v. Paul B. Cole et al., Defendants' motion for partial summary judgment and request for oral argument, p. 4.

11 Ibid., Cole letter Nov. 4, 1976.

12 Gregory, Jim. "Rosboro to reopen" (Springfield *News*, Dec. 10, 1976). Stories in late November in the *Register-Guard* and the Springfield *News* say a total of 127 employees were laid off when plywood end closed down.

13 "Company reopens; 10 employes fired" (Eugene *Register-Guard*, Dec. 13, 1976).

14 Ibid., Gregory.

15 Letter from Paul Cole to employees and families, Nov. 22, 1976.

16 "Lumber company must pay employee fired in dispute" (Eugene *Register-Guard*, March 17, 1978).

17 Ibid., defendants' motion for partial summary judgment, p. 12.

18 The United States District Court for the District of Oregon, Local 2750 Lumber and Sawmill Workers' Union v. Paul B. Cole et al., no. 76-980, Nov. 11, 1976.

19 Ibid., defendants' motion for partial summary judgment, p. 16.

20 "Lumber company must pay employee fired in dispute" (Eugene *Register-Guard*, March 17, 1978).

21 Interview with Paul Cole, late November 2005.

22 Paul Cole comments on Chapter 6 draft, July 26, 2006.

23 Paul Cole, "Company Comment," The *Newsboro*, Oct.-Nov. 1978.

24 The *Newsboro*, March-April 1980.

25 The *Newsboro*, Oct.-Nov. 1978.

26 Interview with Paul Cole, and ibid., Cole comments on Chapter 6 draft, July 26, 2006.

27 Interview with Paul Cole, late November 2005.

28 Cole said the union leaders at that time were "a very difficult group to deal with." They called too many meetings about trivial matters, he said, and they defended workers who in Cole's opinion were doing a poor job. After the strike, he said, the entire union leadership was removed from office. Cole comments on Chapter 6 draft, July 26, 2006.

29 Ibid., Cole comments on Chapter 6 draft, July 26, 2006.

30 The *Newsboro*, June-July 1979.

31 The *Newsboro*, Feb.-Mar. 1979.

32 Michael Thoele comments on manuscript, June 2007.

33 The *Newsboro*, Mar.-April 1980.

34 The *Newsboro*, Feb.-Mar. 1979.

35 Interview with David Weza, March 28, 2006.

36 Interview with Jim Enright, Dallas Davis, and Knox Marshall, Feb. 7, 2006.

37 State of Oregon. Statistics on employment, timber harvest, and lumber and plywood production from 1945-2004.

38 The *Newsboro*, Aug.-Sept. 1977.

39 Burwell, Dave. Article in The *Newsboro*, June-July 1978.

40 "1973 oil crisis" (Wikipedia online encyclopedia, http://en.wikipedia.org/wiki/1973_oil_crisis#Origins_of_the_1973_world_oil_shock), consulted April 28, 2006.

41 McKeever, David B., and Gary W. Meyer. *The Softwood Plywood Industry in the United States 1965-1982* (Madison, WI: USDA Forest Service Forest Products Laboratory, Resource Bulletin FPL 13, April 1984).

42 Paul Cole remarks, the *Newsboro*, May-June 1980.

43 "Softwood plywood production—mil. sq. ft." Table of plywood production nationwide between 1947 and 1981. From Paul Ehinger.

44 Mount St. Helens in Washington, which erupted on May 18, 1980.

45 *Random Lengths* 1981 yearbook (Eugene, OR: Random Lengths, 1981), p. 188.

46 Paul Cole's "Company Comment" in the *Newsboro*, Dec.-Jan. 1978.

47 Paul Cole, quoted in the *Newsboro*, May-June 1980.

48 Although who knows for sure? Maybe the cycle will come around again.

49 Ibid., Cole, *Newsboro*, May-June 1980.

50 Ibid., Cole, *Newsboro*, May-June 1980.

51 Ibid., Cole, *Newsboro*, May-June 1980.

52 Interview with Gordon Culbertson, Aug. 31, 2005.

53 President Ronald Reagan's statement on signing the federal timber contract payment modification act, Oct. 16, 1984 (http://www.reagan.utexas.edu/archives/speeches/1984/101684f.htm), consulted April 28, 2006.

54 Tom Parton, American Forest Resource Council, personal communication, May 14, 2007; he consulted Paul Ehinger.

55 Robert Wolf. Oral history taken by David Jackson, April 18, 1989 (http://depts.washington.edu/wrccrf/Wolf/OH227-04.html), consulted April 28, 2006. Wolf is the author of "National Forest timber sales and the legacy of Gifford Pinchot: Managing a forest and making it pay" (*University of Colorado Law Review* 60(4) (1989): 1037-1078), cited in Gorte, Ross W., "Below-cost timber sales: an overview" (CRS Report for Congress, July 21, 2004), p. 4. (http://ncseonline.org/NLE/CRSreports/04Jul/RL32485.pdf), consulted April 28, 2006.

56 Chadd, Edward A. "Manifest subsidy: how Congress pays industry—with federal tax dollars—to destroy and deplete the nation's natural resources" (*Common Cause National Magazine*, Fall 1995; http://www.ccsi.com/~comcause/news/chadd.html), consulted April 28, 2006.

57 Wolf, Robert. Oral history taken by Dick Manning and Dan Hall, Nov. 17, 1989 (http://depts.washington.edu/wrccrf/Wolf/OH227-40.html), consulted April 28, 2006.

58 ECONorthwest. "Protecting roadless areas and Idaho's economy: an assessment of the Forest Service roadless initiative" (The Wilderness Society, March 2001 (http://www.wilderness.org/Library/Documents/upload/Protecting-Roadless-Areas-and-Idaho-s-Economy.pdf), consulted April 28, 2006.

59 Gorte, Ross W. "Below-cost timber sales: an overview" (CRS Report for Congress, July 21, 2004), pp. 4-5 (http://depts.washington.edu/wrccrf/Wolf/OH227-04.html), consulted April 28, 2006.

60 The *Newsboro*, Dec.-Jan. 1981.

61 But not its forest land, which he kept for himself (Reidel, Carl. "Whose woods are these?" *American Forests*, May 1990). The Diamond takeover enabled Goldsmith to double his money in two years.

62 Ibid., Culbertson interview.

63 The *Newsboro*, Dec. 1984.

64 Interviews with Ruth Cole Kainen and Jim Watts, Aug. 10, 2005.

65 Ibid., Weza interview, March 28, 2006.

66 David Weza comments on Chapter 7 draft, Oct. 27, 2006.

67 Strycker, Lisa. "Owls a symbol in great debate." Eugene *Register-Guard*, April 9, 1989.

68 Letter from Laurence E. Thorp, attorney for Rosboro, to Mel Clausen, BLM district director, Dec. 31, 1986.

69 Letter from Chuck Ivey, chairman of Lane County Board of Commissioners, "To whom it may concern," Feb. 27, 1986.

70 Paul Cole, notes from phone conversation with Ed Kenyon, Dec. 19, 1886.

71 Letter from Richard Eymann to Gordon Culbertson, undated.

72 Gordon Culbertson, notes from phone conversation with Hal Hushbeck, Dec. 18, 1986.

73 Gordon Culbertson, memo to company managers outlining plans for company response to protest, Jan. 5, 1987.

74 Bishop, Bill. "Logging begins at Bunker Hill site" (Eugene *Register-Guard*, date not noted on clipping but clearly early January 1987). There is no record of whether protesters showed up at the office on Jan. 2.

75 Dietrich, Bill. "The owl: bird is a symbol in a debate over survival of wildlife and preservation of the Northwest's timber industry." The Seattle *Times*, May 1, 1989. The words "ecological thermometer" are Dietrich's.

76 Ibid., Strycker, "Owls."

7

ROSBORO STRUGGLES AND ADJUSTS

"It was a hard time for us. A good time, too, but we didn't know that yet."

—David Weza

The listing of the northern spotted owl as a threatened species hit the Northwest timber industry "like a tsunami," said Rosboro CEO David Weza, who at the time was one of two assistant operations managers. The ruling, he said, "was devastating to this valley and to the timber industry."

The industry had only just begun to pull out of the nosedive of the early 1980s. Even before the formal listing of the owl, companies were threatening curtailments, blaming the reduced federal timber supply. The continuing brisk market in Japan for old growth logs was also contributing to the timber scarcity.

The Japanese had been buying logs from Northwest forests for some thirty years, but a strong Japanese economy at the end of the 1980s, coupled with the shrinking federal timber supply, propelled the log-export issue into front-page headlines. Log exports were unpopular with most of the public—the way people in the Northwest saw it, exporting logs was pretty much the same as exporting jobs.

A temporary ban on exporting logs from federal forests had been in place since 1973, renewed every year in congressional appropriations bills.[1] (The overwhelming "Yes" response to an Oregon ballot measure banning log exports from state-owned forests gives an idea of the public distaste for the practice, but the measure was little more than a symbolic gesture, because state-owned forests are only a tiny minority of Oregon's timberland. Moreover, the law did not take effect because of an earlier Supreme Court ruling that states couldn't interfere with interstate commerce.[2])

Private companies, however, were free to market the logs from their own lands in any way they chose. The export market was profitable—foreign buyers were offering a premium of $100 or more per thousand board feet over the roughly $350 that domestic mills were paying, according to one industry expert.[3] And although they weren't supposed to, some companies were taking advantage of loopholes in the rules to export logs from their own lands and then buy national forest or BLM timber to supply their domestic operations.[4]

Courtesy of Gordon Culbertson

Paul Cole, Curtis Ferrell, Larry Taylor, and Gordon Culbertson, 1988

President Ronald Reagan's 1989-90 proposed budget called for lifting the export ban on logs from federal forests. Some large landowning companies were in favor of lifting the ban,[5] but smaller companies were generally opposed. Lifting the ban "would be devastating," said Paul Cole in a newspaper interview, "because we can't pay the prices exporters can pay." Spokesmen for Rosboro competitors Bohemia, Inc. of Cottage Grove and Seneca Sawmill of Eugene agreed.[6]

Members of Congress from Oregon and Washington were also attacking the problem from another angle—working with the federal officials in the new Bush administration to pry open the Japanese market for finished forest products from the Northwest. They issued a challenge to Japanese tariffs on manufactured products, strict building codes, subsidies of domestic wood producers, and other restrictive practices.[7] Subsequent loosening of some of these restrictions opened the Japanese market to American products, especially glue-laminated beams.

In May of 1989 Georgia-Pacific announced it would shut its plywood plant at Springfield permanently.[8] Four years previously the company had sold most of its remaining Lane County timberland (which it had acquired in a hostile takeover of the Booth-Kelly Lumber Company in 1959) to Weyerhaeuser.[9] The mill was eventually purchased by Dick Baldwin, who had run a northern California lumber company, and was reopened as Springfield Forest Products.[10]

Also in May, Weyerhaeuser announced plans to close its Springfield sawmill and lay off its 625 employees, citing difficulties in obtaining federal logs.[11] The company did close, and it tore down its sawmill buildings.[12] The Pope and Talbot sawmill and veneer plant at Oakridge were purchased by a Creswell company called Bald Knob Land and Timber.[13] Within two months the new owners laid off 190 employees.[14] A few of them were later hired back.[15]

Rosboro was doing its best to weather the tough times. In April of 1989 Paul Cole announced possible curtailments in all plants.[16] The late 1980s were "a struggling time for Rosboro," said David Weza. "It was a time when the people of Paul's and Ira's generation were getting ready to retire, a time when a lot of companies were going out of business. Markets were crazy—we didn't know what to sell into."

Updating the Plants

While the whirlwind of change outside might have tempted Rosboro's owners and managers to sit tight, the company continued to invest in needed plant improvements. Here Paul Cole's long habit of meticulous planning and judicious risk-taking stood Rosboro in good stead.

Mill B, the small-log sawmill, was only sixteen years old, but it already needed upgrading. Different ways to saw lumber from small logs were evolving, and computer technology was beginning to revolutionize wood products manufacturing.

The original Mill B had come online in 1975. Its machinery consisted of a quad-band headrig fed with a chain and a 6-inch double-sided edger (boards on one side, cant on the other). This machine was set up with angled rolls that directed the pieces to the right side to be sawed. From the edger, material was transferred directly to an existing planer, which fed to a pull chain left over from the planing-mill days. At the end of the pull chain was a precision trimmer and stacker where loads were banded for shipment.

Paul Cole and operations manager Ira Lowe set out to design a complete renovation. Lowe created a unique double (one over, one under) carriage, loaded by an in-house charger called "the crab." Project engineers under Lowe were Terry Moore and, under him, Roger Liljeberg.[17] Liljeberg and Lowe had worked together at the Feather River sawmill, creating a small-log sawmill there, engineering better flows from the sorter, and adding a lumber stacker.

At the end of two difficult years, the retooled Mill B was up and running, milling mostly Douglas-fir and hemlock logs between 5.5 and 12 inches in diameter. The logs entered a twin headrig consisting of two carriages, one above the other, to make sawing more continuous. The carriages passed between one twin-band mill that fed two 2-inch board edgers and one 4-inch gang saw. The lumber flowed on to the planer, which fed a grading station, and onward to a multiple trim saw, and finally to a pull chain and precision-trim stacking unit.

The goal of the retooling was to achieve higher production and better recovery with about the same manpower, and to utilize smaller logs, which then were cheaper than old-growth logs by about $100 per thousand board feet. In addition, scaling rules for smaller logs made it possible to achieve higher overrun (the term for the percentage of usable wood over and above the standard scale that can be captured from a given log's volume).

Mill B received another sweeping overhaul in 2006. "We went back and replaced some of the equipment," said Cole. "We added a new 6-inch cant gang saw in place of the 4-inch unit. We replaced one of the board edgers with a lineal scanning edger. We changed all of the transfers from the double horizontal resaw to the edgers. We completely redid the headrig scanning and charging of logs to the carriages," and installed an automatic planing mill sorting and grading system. Mill B had previously received an automatic sorting system for the green end of the sawmill, which allowed sawing and sorting of 8-, 9-, and 10-foot lengths, and enabled the drying of individual lengths of lumber, which increased dry-kiln capacity.

The glulam plant also received a major upgrading. When Rosboro's glulam division was started in 1963, manufacturing was mostly a manual process of gluing the laminations, clamping them into a form, and letting the glue dry for eight hours or sometimes longer, depending on temperature conditions. That meant the production of one batch had to be spread over two shifts—the day crew would glue the beams and clamp them in the forms, and the swing or graveyard crew would remove them.

Roof beams

In 1975 the company had invested in a radio-frequency press, which laid up and glued the beams in a much faster, almost continuous-flow operation. The 25-year-old radio-frequency press was still the workhorse of the glulam operation in 1990. It was best suited to manufacturing multiple members of the same size, said Jon Howard, product services manager of the glulam operation. "The primary custom market we were engaged in was the 'roof-cap' market—that is, roof-carrying members for large warehouses in Oregon, California, Arizona, and Washington. These roofs included multiple members of the same size, which fit the capability of the RF press."

In the mid-1980s, however, the company discontinued the custom-beam trade and began manufacturing stock beams exclusively. Rather than being custom-manufactured for a particular application, a stock beam was manufactured to a specified length, such as 60 feet, and entered the wholesale distribution chain that way. The retailer would carry different widths and depths of beams and cut off whatever length the customer needed, much as a fabric store sells lengths of yardage from a bolt.[18]

It was a bold and risky move for Rosboro and a significant change for the glulam industry as a whole, because it required cultivating an entirely different customer base, according to Fred Bauer, then sales manager for the glulam operation, and, later, general sales manager for Rosboro. "We had to teach [the retail yards] what products to sell and when to sell and what not to sell," Bauer said, at the same time that he and his industry counterparts were convincing customers—"indoctrinating" them, as Bauer puts it—of the virtues of the more expensive glulam beams over solid-sawn beams generally. With the dwindling supply of old-growth logs, solid timbers were getting more expensive, and that helped opened the market for glulam, Bauer said.

However, he added, the market for custom beams had become "cutthroat and not so profitable." The reason, said Jon Howard, was that "the western glulam manufacturing industry for some time had been literally controlled by the large roof-cap industry. The vast majority of our volume went to three customers, who dictated prices, production scheduling, and when to breathe."[19]

As a result, "we found ourselves in an environment, much of our own doing, which made it virtually impossible to make a profit," Howard said. "I was involved in the front-end discussions regarding the pros and cons of such dramatic change, and I was intensely involved in implementing the change at the plant level once the decision was made." The decision to move away from custom beams rested with Ira Lowe and Paul Cole. "I can only surmise the process," Howard said. "I know we were tired of losing money and being held hostage by the roof-cappers."

The move was another expression of what was becoming a Rosboro philosophy: find a niche and fill it well. "It proved to be a decision of great insight and foresight," said Howard. By the mid-1990s glulams were being edged out of the roof-cap market by steel. The Japanese market provided a temporary reprieve, but several glulam beam manufacturers subsequently closed their doors permanently, including the two plants that at that time were the largest, Weyerhaeuser's at Cottage Grove and the Willamette Industries plant at Saginaw.

Rosboro's custom glue line was shut down and replaced a few years later by a smaller, less labor-intensive line. The workforce in the glulam plant and office was pared down as well. Production volume was reduced until the second shift on the radio-frequency press came up to speed. "It was a steep learning curve for all of us as we transformed into a 100-percent stock beam producer." A few years later, Rosboro's competitors followed its lead and increased their production of stock beams.[20]

Jon Howard, who came on board at Rosboro at a fairly young age, is among the first of the company's third-generation crop of leaders. Howard was hired by glulam manager Clyde Hughes in 1976 as the plant's finish supervisor. He interviewed on a Saturday, he recalls. "We toured the plant, and then Clyde took me to lunch at the Lucky Woodsman Café. During lunch Clyde offered me the job and I accepted."

Then at the counter, ready to leave, Clyde discovered he had no money with him. "So I had to buy lunch," said Howard. "Thus my Rosboro career was launched."

Within a couple of months Howard was acting superintendent of the glulam plant. "Clyde's background was sales, so he was happy for the help." Eventually Howard succeeded then-superintendent Frank Legner. After Hughes left in 1980, Bryce Weissert, Rosboro's controller, managed the glulam plant for a brief time, and then Ira Lowe took over as manager.

In 1990 Jon Howard left the company to take a job with Bohemia, one of Rosboro's chief competitors, at its newly opened glulam plant at Vaughn, near Noti. Howard wasn't looking for a job, but when Bohemia approached him, he accepted the offer partly because "Ira Lowe and I weren't seeing eye to eye."

In 1993, while Howard was working for Bohemia, Weyerhaeuser made the decision to exit the glulam business and closed its plant in Cottage Grove. David Weza, by then in charge of the glulam plant, negotiated with Weyerhaeuser to buy some of its equipment, including the finger-joint line and planer. Also for sale was a glue-laminated beam press manufactured by a German company called Dimter. Weza was inclined to pass up the press, said Fred Bauer, "but I kept thinking about it." Bauer was convinced the press would be an excellent investment, "so I said, 'David, let's buy it.'"

The Dimter press used the conventional heat and pressure method, rather than radio waves, to lay up and cure the beams. It was a precision machine, turning out beams of higher quality than any other method. "The Dimter press makes basically a perfect beam," said Paul Cole.

Weza and Cole hired an independent engineer named Russ McGuire to supervise installation of the Dimter press. McGuire had been an employee of Weyerhaeuser and knew his glulam equipment, said Weza.

Rosboro's new press was the largest Dimter press in the world. It could make a glulam beam 60 feet long. The company needed to make beams that were 72 feet long, and it needed wider beams, 2-by-8 as well as 2-by-6. Fred Bauer, who speaks German, contacted the Dimter company and asked them about engineering an extension to the press. The Dimter engineers "insisted it couldn't be accomplished," said David Weza. "So we engineered it ourselves, to the amazement of the Germans."

Jon Howard then decided to come back to Rosboro, which suited David Weza fine. Howard's help was vital in fixing start-up problems on the Dimter press, most notably the fact that it wasn't pressing the beams uniformly. Said Weza: "Jon is the best technical person in the industry, period."

The problems centered around getting the new press to turn out beams to the necessary specifications. All U.S.-manufactured glue-laminated beams are certified by the American National Standards Institute (ANSI) as having certain strength and stiffness standards. The certifying protocol takes into account the type of press used to manufacture the beams—presses need to be "qualified" for all sizes of beams they produce.

Rosboro's new Dimter press was not qualified on two of the sizes Rosboro needed to make. Jon Howard was assigned to run tests on the press until the company could prove to ANSI that it could consistently produce beams of the proper strength and stiffness in the two sizes in question. After several weeks the problems were solved, and the new press went into full-time operation. Production volume from the Dimter press initially was small, but subsequent improvements to it eventually made the radio-frequency press unnecessary, and the older machine was shut down in the summer of 2002.

It took a long time for the glulam plant to pay for its expensive renovations, partly because of an ill-fated move into a new high-strength product line called Redi Lam II, which was manufactured with mostly glulam lumber and a few layers of LVL lumber on the top and bottom of the beam. One of its ingredients was laminated-veneer lumber (LVL), purchased from Louisiana-Pacific. The LVL eventually proved defective, "which killed our new product very quickly," said David Weza.

Adds Howard, a touch more philosophically, "Redi Lam II was a little bit like one of Thomas Edison's 5000 failures in developing the light bulb. The failure of Redi Lam II moved us one step closer to developing a high-strength glulam that did work. We, and therefore the industry, learned a lot."

As a result of problems with Louisiana-Pacific's LVL, industry standards for testing structural composite lumber (LVL) in glulam were modified. "Rosboro, innocently enough, had become a training facility for the industry, much to our expense."

Nevertheless, business picked up. In 1996 Rosboro manufactured 40 million board feet of glulam beams, up from 32 million the previous year. Much of its output was going to the then-brisk Japanese market.

In the plywood plant, new equipment added over the years included three new veneer dryers, three hot presses, a high-speed six-belt sander, patch lines, and an automated lay-up line, which replaced the older glue spreaders and allowed the sheets of plywood to be laid up assembly-line fashion.

The company made use of a linear programming model to identify the grades and thicknesses of plywood that would be most profitable. Using the linear program, the company was able to assess the cost of all the types of plywood it was capable of producing, assess the market to determine how much of a given type the plant might produce, and figure how to produce each type at the lowest cost.

For a while the plant's principal product was ¼-inch A/C plywood. Critical to that choice were Raiman patching machines, which cut out knots in the veneer and patched the hole with a clear piece. One-quarter-inch A/C was an item that few plywood plants wanted to manufacture, said Jon Howard, since it requires proportionately more high-grade veneer than thicker types. Thin plywood, obviously, has more of its total wood exposed than thick plywood. A/C plywood requires A-grade veneer for the top side of the sheet and C-grade veneer for the bottom. A-grade veneer, the highest grade, had to be almost clear, with very few knots and defects.

Thanks to the Raiman patching machines, "we got our A faces by using C- and D-grade veneer and patching the knots," said Paul Cole. "We made a silk purse from a sow's ear, in other words." At one time, he said, Rosboro made one-third of all the ¼-inch A/C sanded plywood purchased in the United States.

However, the market was slumping in the early 1990s, and the plywood plant was curtailed in 1994. When it reopened two years later, A-grade veneer was even scarcer and more expensive than before, so the plant was retooled to produce ½-inch CDX plywood.

"Then another thing happened that made it impossible for us to compete," said Cole. "We were using CD veneer and patching it as aggressively as we could for faces, and still make the grade requirements. This meant a lot of patches on the sheets of the panels." Other plywood manufacturers tried to copy Rosboro's success, he said, but with fewer patches on the faces. "Then the southern producers got into the act and their veneer was so good some of their ACX panels had only one or two patches. We knew then we were out of the thin AC panel business."

The veneer plant also was being upgraded. The old peeling equipment, geared toward large logs, was too slow. In 1992-1993 the company installed a new rapid-accelerating lathe drive to accommodate the smaller logs that were, by this time, its main raw material.

Other upgrades included installing new scanning and computer-command systems, converting the dual spindles of the high-speed lathe to reduce the core size to 3½ inches, installing a separate core drive (the machine that drives the veneer block from the top front when the log reaches smaller sizes), and converting from an overhead crane–type charger to a pendulum charger, which swings veneer blocks into the lathe. The periodic bottleneck at the clippers was solved by switching from a knife clipper plus a rotary clipper (which uses six trays), to a system that uses only three trays and feeds directly to the rotary clipper.

The improvements paid off almost immediately in increased yield and value, said Paul Cole, and in nineteen months the company had recouped its $1.3-million investment. "What you see here [at the veneer plant] was happening throughout the company," Cole said. "Our philosophy is to keep all plants up-to-date and highly productive."

Building the Timberland Portfolio

In its woods operations, Rosboro was continuing its aggressive purchases of timberland. In 1990, the Oregon timber company WTD Industries closed down several of its wood-products operations and put about 6,600 acres of timberland in Polk County on the market. The land, on Price Creek near Pedee, had been clear-cut and then abandoned half-planted. "We purchased the land and finished the replanting," said Culbertson. "This was a piece that had been abused, but it was very productive timberland. We could see that in forty years it would be a tremendous asset for the company."

A couple of years later Culbertson learned about 1,800 acres of second-growth timberland for sale in the Coast Range foothills near Falls City. The owner was the Oregon State University College of Forestry, which had received the land in a bequest from the woman who had grown up there.

Courtesy of Gordon Culbertson

Felling a big tree on Finn Rock Tree Farm, c. 1992. Lorne Janz is on right.

Katherine Alice (Kaye) Richardson's family had moved to the cut-over parcel in the early 1930s, when she was a little girl. As she grew up, she learned to manage her family's forest by working with her father, Ward Richardson. (Kaye Richardson and her father were pictured in an Extension bulletin of the 1950s titled, without irony, *Men Who Grow Trees.*) She graduated from the University of Oregon, majoring in journalism, and later went on for a master's degree at another college. Apart from her years pursuing her education, her life was spent on the family tree farm. She worked as a lumber broker and continued to manage her forest using the thinning and selective-harvest methods favored by her father.[21]

Kaye Richardson never married and had no children. By the time she died in 1992, her forest was lush and healthy—and valuable. Her bequest, appraised at $12.4 million, was the largest gift the university had ever received and, indeed, the largest single gift to higher education in Oregon until then. (Shortly thereafter, Nike topped the Richardson record with a $23 million gift to the University of Oregon.[22])

Richardson's will specified that the bequest go to the OSU College of Forestry and directed the dean to manage or dispose of it as he saw fit. After much deliberation, Dean George Brown made the controversial decision to put the parcel on the market rather than keep it. The college already owned

Courtesy of Gordon Culbertson

Logging at Tahkenich Lake

more than 11,000 acres of teaching and research forest just a few miles south of the Richardson tract. The new area was remote and hard to patrol and maintain. Brown's decision was bitterly opposed by some College faculty and many community members.[23]

Because companies were still scrambling for timber, there was keen competition for the sale. The appraised price of the parcel did not reflect its true market value. Rosboro made the winning bid—$23.5 million. "We were particularly interested in that tract," said Culbertson, "because of the shortage of federal timber and because it was a very nice tract, well managed over the years.

He knew the Richardson purchase was politically sensitive. "We understood that some members of the family were strong proponents of uneven-aged management, and they weren't pleased that we could clearcut if we chose. Some people expressed concern that we were going to go in and do what WTD did [on its Pedee forest land]. But I told them that's not the company's style." Rosboro harvested the tract in smaller-than-usual clearcuts over several years, with no significant public outcry.

In 1996 Culbertson traded a parcel of old-growth forest in Douglas County for 2,000 acres of younger forest near Tahkenich Lake. He also bought 12,000 acres of very young forest near Rock Creek, up the Umpqua River, from Weyerhaeuser. "I always had envisioned that, after we harvested the second growth on the Finn Rock tree farm, the Rock Creek tract would be ready to harvest. Our computer modeling showed that it would fit in like a puzzle piece."

Courtesy of Gordon Culbertson

Log merchandiser, mid-1990s

In the same year Rosboro acquired land adjacent to the Springfield plants for a secondary log sorting yard, and by 1997 the company had a new central log processing facility, or, in the trade, "log merchandiser." The name implies the value being added to the logs by sorting, bucking, debarking, and preparing them to enter the mill. Rosboro's merchandiser, which prepares logs for both sawmills and the veneer plant, contained two log debarkers and two automated log sawing lines with infeed decks and conveyors. The new merchandiser was expensive but, like most of Rosboro's improvements over the years, it paid for itself by increasing productivity and reducing manpower needs.

The following year Rosboro bought another 11,000 acres of young forest in Lewis County, Washington, from Weyerhaeuser. The company's land purchases through the 1990s brought its timber holdings to about 75,000 acres, stretching from Douglas County in the south to southwestern Washington in the north, and from the eastern foothills of the Coast Range to the western flanks of the Cascades.

Like many other timber companies in the region, during the 1980s Rosboro had bought as much federal timber as it could, enabling it to stretch out the supply on its own lands. But there was still not enough to operate the plants at capacity. As local sources of timber continued to dwindle, the company looked farther afield. "We bought a lot of timber in Washington to sustain our Springfield operations," said Culbertson. "At one time we were one of the larger contract holders with the Washington Department of Natural Resources." The Washington timber lay in the Capitol Forest, southwest of Olympia, and farther south near Chehalis and Mossyrock.

Courtesy of Gordon Culbertson

John Llewellen climbs a big fir near Toledo

These purchases were risky—the far-flung timber was expensive to haul, which meant each purchase had to be carefully penciled out. "We had to calculate the haul distances really closely," Culbertson said. "A truck had to be able to haul a load in one day, with no reloading, or we weren't interested."

The hard times of the early '80s had hit Washington's small mills even harder than Oregon's, and there wasn't much competition for the Washington timber. "Sometimes we'd get it for a couple of cents over the minimum bid," said Culbertson. "Sometimes there'd be no competition at all."

Culbertson bought timber from as far away as Montana and had it shipped to Springfield by rail. He bought a large Forest Service sale on the Malheur National Forest in southeastern Oregon and had it rafted down the Columbia River. "It was a very large project, a helicopter harvest. We hauled the logs to Boardman by truck, put straps around the loads, and then lift the loads off and put them in the water" and made rafts of the bundles. The rafts were towed downriver and stowed in the backwaters of the Columbia upriver from Portland. When winter came, the logs were pulled out of the water at Troutdale, loaded onto trucks, and hauled to Springfield. "I think we got 7 or 8 million board feet of logs from that sale," said Culbertson. "At that time that was close to a month's production at Rosboro. It turned out to be a really good purchase." Cole recalled that "we even tried buying some spruce logs up in Alaska, which were hauled on a barge called the *Bismarck*."

Courtesy of Gordon Culbertson

Raft of eastern Oregon logs on the Columbia River; Government Island is in the background

At the Finn Rock tree farm, the young trees Dave Burwell had planted and tended in the 1950s and '60s were getting almost big enough to harvest. Burwell had pioneered pre-commercial thinning and reforestation at Rosboro at a time when the idea of turning cut-over land into another forest was still a novelty. The company's intent from the start was to cultivate a forest of stands of staggered ages, making possible a rolling schedule of harvest and a steady supply of raw material for the long haul. In 1990 that vision was becoming reality. "At Finn Rock, we were harvesting old growth at the same time that we were thinning our second growth," said Gordon Culbertson. "We had a full rotation on our forest, and that was unique."

Courtesy of Gordon Culbertson

Logging the Deathball unit, Finn Rock Tree Farm, 1991

Feather River Closed

In 1992, after six struggling years, Paul Cole and the rest of Rosboro's owners decided to pull the plug on its subsidiary in California, the Feather River Forest Products Company. The federal timber the company had been relying on was drying up, and, as in Oregon, logs from private landowners were expensive and ever harder to get.

Moreover, to everyone's surprise, the second shut-down sawmill in Oroville, the one that Rosboro chose not to make an offer on, had been bought by another company, Cornett Lumber. Feather River had a competitor for timber virtually next door. The Cornett sawmill was not a large producer, said Paul Cole, but it influenced the log market significantly.

"Actually, we were surrounded by competitors" for raw material, he said. "On the south there was a big Bohemia sawmill at Lincoln and a closer one, Erickson Lumber, five miles from our mill. On the north was a large sawmill owned by Louisiana-Pacific." Another company, Red Emerson's Sierra Pacific, "was extremely competitive when it came to purchasing Forest Service timber, which we needed badly," Cole said. "They would top any bid we put in, regardless of how high the bidding had gone."

After three years, several million dollars, and countless management hours, Feather River was approaching profitability, thanks largely to the mechanical prowess of Larry Works, Feather River's plant manager. "But it had become very clear that there was no future for us down there," said Gordon Culbertson. "We had a good group—good hourly employees, a good supervisor, a good timber buyer. We did everything possible, but we just did not have the raw material we needed."

Courtesy of Gordon Culbertson

The Feather River mill after it closed

To make matters worse, Feather River Forest Products had become embroiled in a labor dispute. Although Rosboro's Springfield operations had been unionized from the start, Feather River was begun as a non-union operation, and efforts by the employees to organize were met with management resistance.[24] The Lumber and Sawmill Workers' Union filed a complaint alleging that the company was hindering the employees' efforts to organize. The issue ended up in front of a hearings officer from the National Labor

Relations Board. The NLRB ruled against the company, and Feather River was ordered to pay restitution to certain employees amounting to about $50,000.

After the machinery was auctioned off, the plant buildings were razed. "They took them down to the ground," said Culbertson. "It was a terrible sight." The timber still under contract was logged and hauled to the abandoned mill site for transshipment by rail to Springfield. The last employee at Feather River, Russ Lee, stayed until the last truck delivered the last load of logs. Then he unloaded the trucks, loaded the rail cars, and departed for Springfield.

The Feather River episode is not one of Rosboro's finest hours, David Weza said. "We don't like to dwell on it. A lot of blood was spilled down there. It seemed to bring out the worst in all of us."

Even so, he conceded that there were many positive outcomes. "We hired many good people, some of whom transferred to Rosboro [Springfield operations] from Feather River," including Russ Lee. "It was a huge learning and development process, even though it challenged our self-confidence to the core."

Paul Cole also counts the Feather River experience as a net positive. "In a lot of ways Feather River was a fun operation," he said. "We took a run-down mill and upgraded it, adding sorting improvements, a better planing operation, a much-improved log yard. It had a unique small-log sawmill that worked well."

Management Shuffles

In addition to the turmoil in the marketplace and the legal arena, the company was undergoing periodic management shuffles, continuing a pattern that had begun in the previous decade. "It was a hard time for us," said David Weza. "A good time, too," because it led to a more open, more efficient way of running the company, "but we didn't know that yet."

The thirteen years between 1980, when Clyde Hughes left, and 1993, when Ira Lowe left, was a particularly turbulent time, with managers departing under less than happy circumstances and junior employees quitting because they couldn't work with the people who supervised them. At the time the company's divisions were like a group of little kingdoms—"silos," is how David Weza puts it—each with its own fairly autonomous manager.

The glulam department was particularly known for its independence from top-management oversight. It was almost a separate subsidiary of Rosboro, with its own office and a sales staff headed by the colorful and talented Fred Bauer. That was the way Clyde Hughes had set it up when he arrived in 1964, and that was the way it stayed after Paul Cole let Hughes go in 1980. "The glulam plant," Bauer recalls, "had gotten itself into a position where it felt it didn't have any owner."

But shortly after Hughes' departure Cole moved the glulam office staff to the main office—literally, said Jim Walsh. "A house mover picked the building up and connected it to the main office building."

It was "the fall of the kingdom," in the words of Jon Howard. The decision did not sit well with some glulam staff. "It was hard to move back into the general office," said long-time employee Irene Batey, now retired, who'd been recruited from the front office by Clyde Hughes and worked in the glulam office at the time of the move. "The perception was that the lam plant crew was a clique. [When the operation was moved,] we were told, 'you work for Rosboro, not for the lam plant.'" Bauer had a different take: "I was fresh there, so it didn't affect me. From my perspective it wasn't a bad move."

Succeeding Hughes as glulam plant manager for a short time was controller Bryce Weissert, who departed in the early 1980s. He was followed in the controller's job by Mike McCrady. Also about that time glulam sales manager Fred Bauer left the company for a job in Texas.

At that point Ira Lowe was given direct charge of the glulam plant, in addition to his responsibilities for Mills A and B and the plywood operation. After a few years Lowe hired Bauer back. "Fred had gone to Texas and been successful there, but he wanted to return to Oregon," remembers David Weza. In fact, Bauer called Jon Howard and asked him straight-out if there was an opening for a sales manager. "I called at the right time," Bauer said. At the time of his departure Bauer had been in charge of glulam sales. He came back to the larger job of general sales manager for the company. True to Rosboro form, Bauer was no shrinking violet, and he and Ira Lowe butted heads more than once. "Ira would get frustrated with him," Weza said, "and several times he told me, 'If he wasn't so good I'd fire him!'"

Don Cooley had managed the plywood and veneer plants since 1968, when Ira Lowe hired him to succeed Cleve Ramsey.[25] Cooley was a tall, personable man who'd played basketball at the University of Washington. He had worked with Ira Lowe at Weyerhaeuser and was experienced with plywood.

In 1980, Cole had made the decision to put plywood and veneer under separate managers. His reason, he said, was to keep the process "more honest." A single manager responsible for both operations might feel pressured to feed the plywood operation with in-house veneer in order to maintain production, even if the veneer might fetch a higher price elsewhere. Separating the operations made it easier to determine the profitability of each product.

However, in 1982, Cooley was demoted to plywood superintendent, reporting to Ira Lowe. Lowe then took over management of the veneer plant as well, and Cooley was asked to leave.

"I was very fond of Don Cooley—a nice-appearing man, got along well with the people who worked for him, and I thought he knew what he was doing," said Louise McQuin, Rosboro's plywood saleswoman at the time. "I don't know why he left." The reason, said Cole, was "the same reason other managers had to be replaced: they couldn't get profitable results."

Dave Burwell, Rosboro's long-time forester, also departed in 1982.[26] Cole explains: "Burwell had begun as the company's forester and eventually was put in charge of all timber activities. At some point Gene Scherffius was given the job of coordinator, which meant he was directed to get more planning and direction into the woods activities, which we critically needed. At this point, David Burwell was back to mostly forestry-type work. He decided to take early retirement at age sixty-two." Burwell was succeeded by Bob Rand, and then by Gordon Culbertson, who was hired in 1983. Scherffius also departed in 1982.[27]

As manager after manager left between 1980 and 1983, more and more authority accrued to Ira Lowe. "It was like the gunfight at the O.K. Corral," said Jon Howard. "When the smoke cleared, Ira was the only manager left standing." When David Weza came in 1980, Lowe was managing both sawmills and the veneer operation. Two years later he was also in charge of the plywood plant and the glulam plant.

Shortly afterward Cole appointed him operations manager, not only for all Rosboro's manufacturing operations but also for sales. "The ops manager is responsible for profit, and sales are a huge part of that," said Cole. "It's only fair that the ops manager should also be in charge of sales." Lowe's assistant managers were Jon Howard, then superintendent in the glulam plant; David Weza, superintendent of both sawmills and the veneer plant; and Mike Morrissey, sales manager.

An Heir Apparent

By 1983, after the management shakeup was complete and things had settled down a bit, it became clear that Dave Weza's star was on the rise. Compared to the other managers, he was a fairly recent arrival, having been recruited by Rosboro in 1980 to fill a foreman's position in Mill B.

Trained as a science teacher, Weza was recruited from International Paper Co. in Gardiner, where he was working as a manufacturing and process control analyst. Weza was young, only 29, and he looked younger than he was. Arriving early to his first interview, he noticed some nice cars in the employee parking lot—a Mercedes and a couple of BMWs. Then a new red Jaguar pulled in. Its driver was Ira Lowe, he soon found out.

Lowe called Weza into his office and expressed mock surprise at his youthful appearance. "Son, let me see your driver's license," he said. The two men hit it off immediately. "That night Ira called and offered me the job," said Weza.

In his first week on the job he met one of the Arkansas old-timers, Howard Arnold, who would become an indispensable right-hand man to the new superintendent. Arnold had been with the company since 1951, but he still spoke like an Arkansawyer.

"I'm Ha-ard Ahnold," Weza heard the man say as he extended his hand.

"Ah—Howard?"

"That's right. Ha-ard."

"Howard knew where every stick of lumber was in this place," Weza said.

Weza was not closely involved with the plant renovations that took place during the early '90s, especially the conversion of Mill B to a high-speed stud mill. Being an assistant to Ira Lowe, however, he had the supporting role. "I was in a position to advise yes or no," he said, "but I was not invited to get involved in the machinery side." In these matters Ira Lowe had the final say, with Paul Cole's blessing. And Lowe made it clear that he was in charge.

As Weza gained confidence in his new position, he found he didn't agree with all of Lowe's decisions. The Mill B renovation, for instance, did not go well, in his estimation. Lowe was skilled at customizing machinery to Rosboro's particular needs, and with his Seabee can-do attitude he loved doing it. He was perfectly comfortable taking off-the-shelf machines and processes and tailoring them to Rosboro's needs as he saw them.

Weza thought inventing, tinkering, and custom-rigging was the wrong way to go about creating a state-of-the-art sawmill. To him it was a throwback to the "haywire" operations of the old days. Nonstandard machinery was subject to glitches and breakdowns, and it was a source of constant frustration to the people who had to make it work while it was being invented right underneath them. "My philosophy," Weza said, "is to buy equipment that has a proven track record."

Although the small-log sawmill did finally achieve the expected improvements in efficiency, the remodel took much longer than expected and cost $2.5 million more than it was supposed to, Weza said.

Weza also had doubts about the purchase of the Feather River plant. Not yet in the top echelon of management, "I didn't know anything about that deal until it was done," he said. "In my opinion it was not a good decision. It was not well thought out. We had a vision that wasn't really very good—that we could make a small-log mill out of a large-log mill." But the Feather River purchase had the blessing of Paul Cole and the other owners (although some of them had their doubts, too[28]), and Weza knew enough to keep his objections mostly to himself.[29]

Although they'd gotten on very well in the beginning, Weza and Lowe were starting to butt heads on a more personal level. Lowe's rise in the company was powered by his technical skills, which were formidable. "I respect Ira a great deal, and he taught me a great deal," Weza said, "but the people side of managing was a challenge for him."

By 1983, Weza said, "I was getting pretty much fed up. I was seeing good people leave." He had a heart-to-heart with his grandfather, whom he regarded as a wise mentor. "My grandfather asked me, 'Do you like the people there?' I said I did. He said, 'Then why don't you try to make a difference rather than running away?'"

Weza decided to keep trying. In 1986, out of the blue, Paul Cole came to him and told him that if he was contemplating leaving the company, please come and talk to him, Paul, first.

"I hadn't told Paul I was unhappy," Weza said. "Howard [Arnold] may have told him. He was the kind of guy who could be invisible he could be anyplace." However Cole learned of Weza's frustration, the gesture of support from the top boss was reassuring.

In some ways the differences between Lowe and Weza may have stemmed from a simple generational conflict. Lowe was the World War II veteran, raised with an ethic of hard work, frugality, and self-sufficiency, and he was comfortable with paternalistic, even authoritarian leadership. Weza was the

Baby Boomer, raised in an era of expanding prosperity, full of drive and ambition. He was not about to take "because-I-said-so" lying down.

After several tense months of fussing with the machinery in Mill B, Paul Cole began to lean on Lowe a little harder to get results. At more than one point "we had to shut the plant down, and it took time to fix it," Cole said. "After that, I started pushing him on the engineering stuff. I wanted better planning, better investigation."

Ultimately it became clear that the conflicts between Lowe and others at Rosboro ran deeper than mere generational differences or disagreements over methodology. In 1993, Ira Lowe and Paul Cole had a falling-out. Cole was displeased with the problems in the Mill B renovation. He did not approve of what he considered Lowe's proprietary attitude toward his own mechanical inventions. He was not happy with Lowe's strained relationships with some of his co-workers and subordinates.

The upshot was that Lowe decided to retire. Just before an owners' meeting in San Diego, he tendered his resignation.

By all accounts Ira Lowe left a solid legacy in his twenty-seven years with Rosboro. "He made this company a lot of money," said David Weza—the highest possible praise in a company with such a single-minded focus on profit. Lowe's energy, commitment, creativity, technical wizardry, eagerness to embrace change, and can-do competence all left a positive mark, said Paul Cole. "Ira was a very important person in the growth and diversity of the company. Rosboro is what it is largely because of his contributions."

Lowe now lives in Charleston, Oregon, with his wife, Phyllis. He's an avid deep-sea fisherman and an inveterate remodeler: he spent the past couple of years laying a new hardwood floor in the couple's home and building a snug apartment over the huge shop where he keeps his two boats.

He remembers his years at Rosboro mostly with pride and pleasure, but he acknowledges the difficulties he experienced. "When I started there, I was superintendent of the dimension mill, and that was all," he said. "I only had to run the mill. Then I got the lam plant, then the veneer plant, then plywood. Adjusting to all that was a real challenge. I liked the production part the best. Managing was not something that I had a high feeling for."

Rosboro was good to him, he said. "I never had a problem with Rosboro and I still don't. I had a great partnership with Paul Cole. I have a lot of respect for him and his ability. We were great together."

Lowe was succeeded by Fred Mondou, who had been superintendent in the Rosboro plywood plant. Mondou came with good computer and systems-engineering skills, as well as being a very personable man, David Weza said. "He was a good manager. He could be tough when he needed to, but he worked very well with people." Mondou's tenure was tragically cut short in 1997 when he died from a heart attack at forty-eight.

"I was devastated," said Weza. "Fred was not only manager of all the plants, but a very good friend. It was a hard time." After that Larry Works

Fred Mondou

became interim operations manager. He was followed by two others, Kerry Hart and Gary Thebault, before Dallas Davis, the current operations manager, came to Rosboro from Weyerhaeuser in 2002.

The passing of the baton from Ira Lowe, the hands-on practical engineer, to Fred Mondou, the systems-savvy, people-smart facilitator, is a good emblem of how much the business environment had changed for Rosboro Lumber Company. Everything was getting more complicated all the time—manufacturing systems, markets, politics, and most important, human relations.

"I used to think it was all simple," muses Ira Lowe. "I had things I wanted done. You just think about what's going to happen, and you plan for it, and you do what you need to do.

"Nothing was ever complicated with me. Maybe that was my problem."

Notes

1 Rhodes, Jerry. "Devastation seen if federal log export ban lifted." Springfield *News*, Jan. 11, 1989; and Swisher, Larry, "More log export foes hop on 'ban wagon,'" The Eugene *Register-Guard*, March 13, 1989.

2 "Log export ban wins big." Springfield *News*, July 3, 1989.

3 Ibid., Rhodes, "Devastation seen;" and Robertson, Lance, "Packwood wants forest law changed," Eugene *Register-Guard*, May 24, 1989.

4 Robertson, Lance. "Mortgaging our future?" Eugene *Register-Guard*, July 9, 1989.

5 Although not all—Weyerhaeuser supported keeping it. Swisher, Larry, "More log export foes hop on 'ban wagon,'" Eugene *Register-Guard*, March 13, 1989.

6 Ibid., Robertson, "Mortgaging our future?"

7 The Associated Press. "Barriers to wood products studied," Eugene *Register-Guard*, June 15, 1989.

8 Strycker, Lisa. "Georgia-Pacific closing Springfield plant." Eugene *Register-Guard*, May 10, 1989.

9 Ibid., Strycker, "Georgia-Pacific," and Thoele, Michael, *Bohemia: The Lives and Times of an Oregon Timber Venture* (Portland: Oregon Historical Society), pp. 313-14.

10 The company later bought Triangle Veneer on Highway 99 in Eugene. Springfield Forest Products eventually went out of business. GE Capital Partners bought the plant and reopened it as McKenzie Plywood. It is still in operation as of this writing. Comments from Paul Cole on Chapter 7 draft, November 6, 2006.

11 Rhodes, Jerry. "Log shortages threaten future of mills." Springfield *News*, April 22, 1989.

12 Comments from Paul Cole on Chapter 7 draft, Nov. 6, 2006.

13 Boyd, Jim. "Bald Knob purchases Pope & Talbot mill." Eugene *Register-Guard*, May 23, 1989.

14 Portal, Ann. "Oakridge mill layoff idles 190." Eugene *Register-Guard*, July 28, 1989.

15 Portal, Ann. "Some workers back on job." Eugene *Register-Guard*, July 29, 1989.

16 Rhodes, Jerry. "Log shortages threaten future of mills," Springfield *News*, April 22, 1989.

17 "Ira always had an engineer working for him," said David Weza. "The first one I remember was Earl Jenks, then Dennis Dorrah (who came from Heinz Lumber in Burns), then Doug Wilson, who was Paul Cole's son-in-law at the time. Doug had worked under Dennis and was promoted when Dennis left. Then Gene Anderson was hired (he came from Eugene Sand & Gravel) and Doug Wilson worked under Gene. When Gene left Doug was promoted. He did not last very long. Gene had hired Roger Liljeberg to assist him. Then Terry Moore

was hired and Roger worked for Terry. When Terry was fired Roger assumed the role. Roger left about one year after Ira left the company." David Weza comments on draft.

18 Interview with Fred Bauer, Feb. 6, 2007.

19 Interview with Jon Howard, March 30, 2006.

20 Comments of Jon Howard on draft of Chapter 7, Nov. 6, 2006.

21 Taken from "The Gift of a Forest." *Focus on Forestry*, a publication of the Oregon State University College of Forestry, Fall 1993; interview with George Brown of March 2, 2007; and recollections of the author.

22 Interview with George Brown, March 21, 2007.

23 The proceeds of the Richardson property gave the college a new building, Richardson Hall, dedicated in 1999, as well as three endowed chairs, an endowment to the college's teaching program, and some money left over. Selling the property "was one of my better decisions," said Brown, now dean emeritus, "because it parlayed that gift into something far more valuable."

24 Ibid., Weza interview, Aug. 30, 2005. He said the early managers tried to hide Feather River's affiliation with Rosboro so that the union would not find out that the parent company was a union shop. "But everybody in town knew."

25 Gardner, *Rose Burro*, Nov. 27, 1968.

26 Interview with Louise McQuin, March 30, 2006. Burwell was "crowded out," she said.

27 Scherffius was not asked to leave, he said, "but I got wind of [a] coming demotion to myself and Don Cooley." He submitted his resignation in May of 1982 and then decided to stay on until the fall. Comments on draft 3, April 21, 2007.

28 Ibid, Kainen interview, Aug. 10, 2005, and interview with Jim Watts, Aug. 9, 2005.

29 Opinion remains split about the wisdom of buying the Feather River mill. Gordon Culbertson asserted that the decision was well planned—"We did a lot of due diligence." He and Ira Lowe both viewed the Feather River adventure as a good move plagued by bad luck. "We thought we made a pretty good deal, but then the timber got out of control," Lowe said. Owner Ruth Cole Kainen's verdict is softer than David Weza's: "There have been some decisions I haven't agreed with—the California mill, and sure enough it didn't turn out. But that was because the federal timber was gone." Culbertson interview, Aug. 31, 2005, and Kainen interview, Aug. 10, 2005.

8

INTO THE TWENTY-FIRST CENTURY

> "We set measurable goals at every level. … If you can't measure it, it's not worth doing."
>
> —*Dallas Davis*

By the turn of the century, Rosboro had survived a turbulent decade. Events outside the company had made for an operating environment that was unpredictable, to say the least.

More and more, Oregon wood producers found themselves competing in a global marketplace. Lumber from Canada was pouring over the border, to the complaints of U.S. producers, who maintained that the Canadian government's involvement in the timber industry there amounted to an unfair subsidy. Heavy duties imposed by the United States were appealed by Canada to the World Trade Organization and NAFTA, but the dispute would drag on throughout the decade and into the next century.[1] Imports of lumber would also continue, not only from Canada but from Europe and South America.[2]

The Bubble Bursts

The Japanese markets for U.S. logs and other products, notably glulam beams, collapsed suddenly with the Asian economic slump in 1997. Asia, and Japan in particular, had been enjoying an economic boom since the early 1990s. The 7.3-magnitude earthquake that devastated the cities of Osaka and Kobe in January of 1995 boosted Japan's already-strong home-building industry. Within a few months, planning had begun on a massive reconstruction project that included 110,000 units of new, fire- and earthquake-resistant housing.[3,4] The reconstruction required millions of board feet of lumber and structural wood products, and Rosboro's glulam beams were part of the mix. The opportunity to provide Japan with wood beams for its traditional post-and-beam housing construction helped the company offset competition from the steel industry, which took advantage of rising wood prices in 1993 to move into the American market for structural members.

David and Paul headed for Japan

But then in November of 1997, the Tokyo stock market went into free fall, and most of Asia plummeted into a decade-long recession. Japan's boom had proved a fragile bubble, kept artificially inflated for most of the decade by the major banks' propping up of weak companies, coupled with a tacit expectation among large companies that they should help one another out in hard times, a system known as *keiretsu*.[5] The result was a system of entangling alliances that eventually brought the strong companies down along with the weak ones.[6]

The bursting of the Asian bubble was a wake-up call for the United States and its international partners. So interconnected had the global economy become that the whole world was vulnerable to a downturn in one region.[7]

Rosboro had been shipping half its glulam output to Japan, according to Jon Howard, product services manager for the glulam division, and other U.S. companies had added capacity to meet the Japanese demand. When Rosboro finished rubbing its eyes, "we were surprised at how bad the U.S. markets had become," owing mostly to increasing competition in the commercial roof market from producers of steel beams. "We were in a deep hurt."

Party group poses under Rosboro banner in Tokyo

Moreover, overseas wood products manufacturers had entered the competition for the Japanese market. Rosboro found itself being squeezed out by northern European companies, who combined aggressive pricing with lower shipping costs to undercut U.S. firms, according to Paul Cole. "They took over the market in two or three years," he said. The heavy competition eventually led to the company's decision to abandon the Japanese market in 2003.

Closer to home, mill closures, departures, mergers, and acquisitions among Northwest wood-products companies continued unabated. By the end of 1990, thirty-nine Oregon mills had closed their doors.[8] In 1991 one of Oregon's home-grown companies, Bohemia, Inc., was swallowed by a larger one, Willamette Industries—which would, in 2002, be swallowed by the multinational giant Weyerhaeuser Co.[9] Three years earlier Weyerhaeuser had acquired MacMillan Bloedel, one of Canada's largest wood-products companies, in a $2.45 billion deal.[10]

David Weza and Paul Cole at owners' annual meeting in Vancouver, B.C., September 1999

Change of Command

Rosboro was sixty years old, and its world had changed utterly. Still family-owned and relatively small, Rosboro did not feel the same relentless pressure for short-term profit that larger, publicly traded firms did during those tempestuous times. Nevertheless, the company was coping with the same market conditions: high prices for logs, unpredictable prices for finished goods, volatility in interest rates and housing starts, and competition from overseas suppliers. In August of 2000 Rosboro shut down its dimension mill indefinitely, and in September Mill B was cut from three shifts to two.[11] "Dozens of mills across the country have cut production or closed indefinitely over the past three months as wood products prices continue to slide," reported the Eugene *Register-Guard*.

Paul Cole had been running the company for forty-two years. He was seventy-three years old—older than his father had been when he'd passed the reins to Paul. He was in good health; still active, decisive, sharp. What better time to do some hard thinking about the company's future? Careful planner that he is, Paul Cole began to set in place the mechanisms for others to succeed him.

The company had operated as a partnership for most of its life. In 1996 the Rosboro companies (there were two by then, Rosboro Lumber Company and USR Company) were changed from partnerships into LLCs (limited liability companies). The LLC setup gave Rosboro the corporation's advantage of limited liability, but allowed operations to remain pretty much as they had been under the partnerships. Of course, the owners remained the same.

Peeler blocks in the veneer yard, 1997

David Weza, by then executive vice president, began a comprehensive strategic plan. "We were being dragged in so many directions," said Weza. "The world had changed, and it was obvious that our business was becoming more and more challenging."

In June of 1998 Weza got in touch with John Martz, who had been a consultant for Rosboro for almost twenty years and had helped the company with its earlier leadership training. Did Martz know of anyone who could help the company figure out what to do next? Martz got in touch with Rip Van Camp, a retired R. J. Reynolds executive then living in the San Juan Islands of Washington, and set Weza up with a phone interview.

Weza liked what he heard. He traveled to Seattle and met Van Camp at the Washington Athletic Club. They discussed the company's management, its key players, and its uncertainty about the future. Should Rosboro expand? Diversify? Divest itself of the mills and become a timber-only company? Or go the other way and buy some more mills?

A few days later Van Camp called Weza. "I'd like to help you," he said, "but you're getting ahead of yourself.

"How's that?" asked Weza.

"Your strategic plan—well, it doesn't really exist."

A stock-taking was in order—a complete inventory of the company's goals and strategies, strengths and weaknesses. Lynn Damon, the company's financial analyst, provided the detailed numbers to do a thoroughly quantitative job. Out of this exercise grew a strategic plan with measurable goals. Among other things, the plan called for investing in a comprehensive management information system, identifying the most profitable product lines and abandoning the others, and becoming more choosy in purchasing timber and timberlands.

Most importantly, the plan forced the company to become quantitative in everything it did. "We had to develop measurables and clearly communicate them through the company," said Jim Walsh. "We had to set quantifiable goals, so we could track our progress. Now, if we're off the plan, we see why, and we reassess. These are living plans—they're constantly changing—but everyone on the team knows what the goals are at every moment."

In 2001, with the approval of Rosboro's board of directors, Cole appointed David Weza as chief executive officer. Weza had been at Rosboro for twenty-one years and had risen to second in command at Rosboro—a long journey from his first job as superintendent of Mill B. "I never aspired to be the CEO," he said. "I didn't think I'd be here that long. When I was first hired, I thought I'd stay maybe five years." But Cole kept increasing Weza's challenges, kept giving him more responsibilities, and he was clearly pleased with Weza's performance.

Another account of Weza's rise through the company suggests that he did have his eye on the top post early on. Larry Taylor is Rosboro's contract supervisor and a man who, like Paul Cole, grew up with the company. He can tell some colorful stories, and he tells this one about David Weza:

> One time pretty soon after he was hired, I talked to him at some company function and asked him how it was going. He said, 'pretty good.' I asked him what his plans were, and he said, 'Oh, I'd like to be the CEO some day.' I

> laughed and said to him, 'You cracker-ass, they aren't gonna let you run this company.' Well, the years went by, and what do you know, he's the CEO!

Taylor and Weza met at another company function soon after he was promoted, and Taylor said to him, "Well, cracker-ass, I guess you made it." He said Weza took the comment good-naturedly.[12]

Courtesy of Gordon Culbertson

Loading logs from Plane Crash unit at Finn Rock, 1999. Ray Froppe stands on lowboy

David Weza's promotion to CEO signaled a seismic change: after forty-two years, Paul Cole was no longer in control of Rosboro Lumber Company. Cole had been planning for this event for a long time. As Hub Lindsey later told him, "You have done what many top managers have not done, and that is to train and select a successor and then put him into place to manage the overall company. You should feel proud of how this has worked out."

It cannot have been easy for Cole to give up control of the company that had been the governing presence in his life since he was transplanted from Arkansas to western Oregon as a boy. "Paul has hung on because he's obsessed with that company," said his wife, Nonnie. "That's what it takes, that 100-percent commitment."

David Weza said the word "retire" is seldom used in connection with Cole. "He'll continue on here until he dies." Another expression that several people used in this context is "feet first." Cole continues to lend his counsel to Weza and the other managers, but the decisions are now Weza's to make.

Taking down the Silos

With Weza now at the helm, he and Cole began to talk about reorganizing the management structure. The current management was grouped, logically enough, by operation—timber, lumber, glulam, veneer, and plywood. This scheme tended to turn the plants into "silos," said David Weza, and to foster divisions and competition among managers. A better way was needed—one that kept people's focus on the good of the company as a whole.

Weza wanted to replace these vertical divisions with horizontal ones that cut across all the manufacturing operations. The management would henceforth be organized into "business unit teams," one for each plant and one also for the resource (timber) side. Each team would consist of a representative from each of the company's functional disciplines: accounting, finance, sales, resource, operations, and others as needed. The business unit teams would be overseen by an executive team with David Weza at the head.

In 1995, as sales manager Fred Bauer was getting ready to retire, David Weza hired Jim Walsh from Willamette Industries to succeed him.[13] It was Bauer, in fact, who brought Walsh's name and accomplishments to David Weza's attention. But after Walsh was on board, Bauer kept putting off his retirement.[14] He finally left in about 2000. As the company's planning progressed, Weza added business team development and ongoing strategic planning to Walsh's duties. Soon after that he gave Walsh the title of business development manager, charged with planning and scouting opportunities for expansion and diversification. No such portfolio had ever existed before. Because these planning activities were taking quite a bit of Walsh's time, Weza hired Jim Enright, also from Willamette Industries, to succeed Walsh as sales manager.

The first executive team, then, consisted of operations manager Gary Thebault, who had succeeded Fred Mondou in 1998; resource vice-president Gordon Culbertson; business development manager Jim Walsh, general sales manager Jim Enright, and financial analyst Lynn Damon.

The new management structure took some getting used to. "It was such a dynamic change," said Weza. "Now we were asking for the walls to come down and for people to talk to one another and to work together. In the beginning people were very quiet in meetings. They weren't sure what their role was."

The new arrangement, with its more open lines of communication, made it harder to hide underperformance, whether from a superior, a subordinate, or oneself. "It's said that cream rises to the top," said Weza. "I believe that. But so do the dead fish." Translation: both good performance and poor performance become obvious in an open environment.

During Weza's first year as CEO, Gordon Culbertson, by then vice president for timberlands, left the company. Neither man will be very specific about why. The trigger seems to have been conflict over the management direction the company was taking, expressed in differences of opinion over the best ways to acquire raw material. Said Cole, "Things were changing in the industry, particularly in timber requirements, and like many timber managers he found this difficult to deal with."

For his part, Culbertson said simply, "It was time for me to do something else. I'm proud of our accomplishments during my years with Rosboro, and I have great respect for Rosboro still."

Indeed, the business team process had turned a spotlight on every link of Rosboro's supply chain, from raw material to finished product. The result was a sharper focus on the coordination between each step in the manufacturing process. No longer could the company afford to buy logs that weren't exactly the right ones, in exactly the right quantities and sorts.

"In my experience in this industry, the connectivity between raw material and end use is disjointed," said senior resource manager Knox Marshall, who stepped into Culbertson's vacant position in 2003. "It wasn't just at Rosboro. In most places, the mill guys and the timber guys aren't used to talking to each other."

Companies may have been able to get away with that when logs were cheap and plentiful, "but now it's essential to maximize the value of the process. We're all accountable to each other" for the efficient flow of raw material through the plants.

"Accountability" is a favorite word of David Weza's. "It wasn't that the company wasn't being managed well" when he took over as CEO. "But I'm a huge believer in accountability, and there were areas where the accountability needed to be improved."

Building on the work of the late Fred Mondou, who as operations manager had installed the company's first management information systems in the plywood plant, Weza invested in a company-wide management information system in early 2002. The system enables managers to monitor operations of each plant almost minute by minute and make the small, crucial decisions that keep processes as efficient as possible.

Lynn Damon, Rosboro's financial analyst and, at the time, head of information technology, played a lead role in guiding the company through the automation process. Damon, who incidentally is the daughter of Ira Lowe, has worked for Rosboro for most of her career. "I was here when we got our first PC [computer]," she said. She led the company in researching the various systems on the market and identifying the product, ScoopSoft, that best fit Rosboro's needs. She worked with ScoopSoft's enterprise resource planning team to install the components of the system and get people trained to use it. "There were people in our plants who had never used a computer before," said Damon.

The project took nine months from start to finish. Much of that time was spent in clarifying the company's needs and talking through the details of how to meet them. "It was very time-consuming," Damon said, "and there was sometimes a feeling of 'let's just get it done.' But it was time well spent. We worked out all the kinks and bugs we could, and then we just went live."[15]

Meanwhile, David Weza and Jim Walsh, with input from consultant Van Camp, were considering the company's prospects for diversification or expansion. In the fast-changing world in which Rosboro now operated, what kinds of opportunities should it pursue? Walsh and Weza, with the help of Van Camp, began developing detailed criteria. Mindful of limited log supplies and the typically low profit margins in commodity products, they decided Rosboro should keep its eye out for wood-based manufacturing companies that added more value to the raw material and promised good long-term profits.

In 2001 Rosboro entered negotiations to buy a small business in Seattle that manufactured hardwood window and door frames. Started by two brothers in the early 1980s, Washington Hardwoods built composite-core hardwood door and windowframes for the Seattle commercial-construction market. Much of its product was going into high-rise office buildings downtown. The company was undermanaged, said Walsh, and he believed its purchase would put the company into a profitable specialty market and into new commercial territory as well. Washington Hardwoods seemed to meet all the criteria. However, the collapse of the dot-com industry followed by the Sept. 11, 2001 attack dealt Seattle's high-tech industries a devastating one-two punch, and the market for Washington Hardwoods' products nearly collapsed.

As the economy picked up after 2004, the company developed new lines of engineered-hardwood moldings, millwork, and doors. The plant has been re-engineered to improve its efficiency, and a new plant manager and a new sales manager were hired. Washington Hardwoods started to turn around in 2005, after a long struggle.

A New Glulam Plant

Midway through the first decade of the new century, the lumber industry was in an unaccustomed bullish mood.[16] A strong, sustained housing market and record low mortgage interest rates boosted demand for wood products, as did the devastating hurricanes of 2004 and 2005.[17]

Rosboro's Mill A, the big dimension mill, idle since 2001, was restarted in 2004 to produce timbers and laminating stock for the glulam plant. The next year, after much research and deliberation among the company's senior management, Rosboro entered negotiations to buy Weyerhaeuser's glue-laminated beam plant at Vaughn, in the Coast Range foothills west of Eugene.

The plant, which sits on 155 acres about 10 miles southwest of Veneta, had been built by Bohemia Inc. in 1988,[18] three years before Bohemia was acquired by Willamette Industries. Jim Walsh and Jon Howard knew the plant intimately—Walsh had been sales manager for Bohemia's glulam division when it was built, and Howard had managed the plant. Jim Enright and Dallas Davis also were familiar with it. Davis had operated it for Bohemia at the time the company was taken over by Willamette, and Enright had headed up the sales team for Willamette's engineered wood products. They all knew the plant was in good physical shape, and they knew most of its employees had been there for a while—the average length of tenure for the employees was fourteen years.

Two of Rosboro's sales staff, Michael Kirkelie and Cindi Hengstler, also had experience with the Vaughn plant and its products. Kirkelie had sold glulam for Bohemia glulam for several years before he came to Rosboro, and Hengstler sold for both Bohemia and Willamette Industries before coming to Rosboro.

In 2002 Weyerhaeuser acquired Willamette Industries in a bitter and drawn-out hostile takeover. With the package came the Vaughn glulam plant. In its other operations, Weyerhaeuser manufactured I-joists, parallel strand lumber, and laminated veneer lumber (LVL) under the trade name Trus-Joist. These products competed aggressively in the same market as Rosboro's glulam beams, said Jim Walsh, and glulam was not considered a core product for Weyerhaeuser.

Mike McCrady

Shortly after Weyerhaeuser's takeover of Willamette Industries, Walsh, David Weza, and Paul Cole met with Rich Hansen, Weyerhaeuser's senior vice president for timberlands, and executive vice president Bill Corbin. Rosboro had no specific agenda, said Walsh—"we just wanted to reintroduce ourselves and try to understand where they were going as a company."

The executives of the two companies met in Portland but "did not have a strong dialogue," in Cole's estimation. But at the end of the meeting, Bill Corbin pulled Walsh aside and asked if Rosboro wanted to take a look at the Vaughn plant, because it did not fit Weyerhaeuser's needs. Rosboro would think about it, Walsh told him. Later Corbin intimated that perhaps he had been too hasty. Weyerhaeuser needed glulam to supply its distribution yards. Maybe they would keep the plant after all.

Over the next couple of years the companies stayed in touch. With the help of operations manager Dallas Davis, who had come to Rosboro in 2002 from Weyerhaeuser, the company looked hard at the plant's operations and future potential. On the finance side, chief financial officer Scott Nelson, with the considerable help of Lynn Damon and Rich Babcock, ran the numbers through every possible contingency.

"This decision was a good example of one of our core strengths: that we make decisions based on fact, not on emotion," said Damon. "We looked thoroughly at every aspect. Sales—most of the existing Vaughn customers were Weyerhaeuser wholesalers. What would happen if we lost half of them? All of them? Who would we replace them with? What presence would we have in our market areas? What level of production would that mean?"

Similar detailed questions were asked about raw material supply. What were the implications if the company had to buy all its glulam stock from outside, or if it manufactured all of its own? Operating and staffing were scrutinized. The Vaughn plant was judged to be overstaffed under Weyerhaeuser, and Dallas Davis foresaw some savings in reducing staff.

Damon gathered facts from everyone concerned, ran a long list of what-if scenarios, and subjected each to what she calls a "reasonableness check." Damon has been at Rosboro for a long time and knows how things work, and the company trusts both her analytical skills and her judgment on what is reasonable. "Besides, I'm outspoken," she said. "If something doesn't seem right, I say so. In this case, it was a very good idea."

And so, said Walsh, "we decided it would fit us very well."[19]

In the summer of 2005, Bill Walters, Weyerhaeuser's business development director, and Vinayak Maheshwari, project director of acquisitions and divestitures, invited Rosboro and two other companies to make an offer on the Vaughn glulam plant. Rosboro's bid won, and in October 2005 the company took possession.[20]

Rosboro wanted the handoff to be short, sweet, and smooth, and it was. There wasn't much time: five days between closing and possession. IT director Don Halstead and his team had to swap out the existing management-information data system and replace it with Rosboro's ScoopSoft, which was completely different. His team had to set up the new phone system, bring in new computers, and train the employees. "He had a very tight time frame to get the system up and running and train people to use it," said Walsh. "He and his team did a fantastic job. All systems were up and running" on the appointed day.

In the course of the financial analysis, Rich Babcock, with the help of Wes Anderson and the human resources team, developed a pay and benefits package comparable to Weyerhaeuser's—a condition that Weyerhaeuser had imposed on the sale. The Vaughn employees, who were non-union, needed to be compensated similarly to the union employees in Rosboro's other operations. The human resources team quickly got everything set up, taking extra care to keep the employees, who were understandably nervous about the change, fully informed.

After the purchase, the new glulam plant went smoothly into operation. The Vaughn crew were ready to show their new bosses just how good they were, said Paul Cole. "In the first full month of operation, the plant increased production by 18 percent, and they never looked back."

The sales team managed to hang onto all Weyerhauser's previous customers, about 85 percent of which were Weyerhaeuser-owned wholesale distribution yards. It was critical that Rosboro retain these customers, but the plant was behind in its orders, and Rosboro couldn't satisfy everyone immediately. Enright and his team forged cordial relationships with all the customers and skillfully allocated what product there was among them. "Jim and his crew worked hard to keep everybody happy," said Walsh. At the end of six months the orders were up to date and shipping smoothly. "Our Weyerhaeuser customers were very complimentary of our plant managers and our sales team," Walsh said.

Early in 2005 Rosboro made the decision to outsource the remainder of its logging and trucking operations. The company had already been hiring contractors to do most of its logging and about half its trucking already. It was one of the last companies of its size that still employed in-house loggers and truck drivers. The decision was driven not by a need to economize, but by a shift in Rosboro's strategic direction, according to resource manager Knox Marshall. Larry Taylor, then logging superintendent, knew it had to happen, but he was still troubled by the decision. "I was their boss," he said, "and I had to break the news to them, and I didn't want anybody else doing it. That was hard. Heck, there'd been guys working for us for twenty years."

For its hauling needs Rosboro contracted with Knutson Trucking, which bought the company's trucks and leased its truck shop. Most of the laid-off Rosboro truckers went to work for Knutson, and most of the loggers found work with various local contractors.

Embracing Excellence

The twenty-first century has been good to Rosboro so far. A more settled operating environment outside the company, and Weza's leadership inside, seem to have produced a stable, enthusiastic, and committed cadre of managers. The latest member of the current executive team is chief financial officer Scott Nelson, who joined Rosboro in 2005.

Sales manager Jim Enright has talked to enough old-timers to sense that Rosboro has departed from the pathways of the past. "The management structure, the working environment, the team dynamics—everything has changed." He said David Weza has helped Rosboro transform its company culture, combining the best of the traditional strengths with new ones.

For example, an unwavering focus on profit has always been the Rosboro way, and finally having an effective management-information system helps keep profit on the top of the list. "David is not an ego-driven manager," said Enright, "and we're not an ego-driven company. David keeps us focused on our primary goal, which is to make money for the owners."

The new management team also pushes research and planning to ever higher levels. As the dimension mill was being started up again in 2003, Dallas Davis worked with Lynn Damon to conduct extensive trials with many logs of different sizes, species, and grades to determine the best mix for the timbers and laminated-beam stock the mill was to produce.

Rosboro's preoccupation with accountability, too, has been taken to new heights. "We set measurable goals at every level," said Dallas Davis. "We have goals, and then we have 'stretch' goals, and we measure ourselves against the numbers that come out every day. If you can't measure it, it's not worth doing."

Finally, said Knox Marshall, there was a philosophical change of gears as the company moved away from a mentality of good-enough to embrace an expectation of excellence. "There was a time, for example," he said, "when Rosboro would accept poor-quality timber. I'm told it got so that Rosboro became a dumping ground for logs that other companies didn't want. Weyerhaeuser had 60 log sorts—Rosboro had one." That began to change, he said, when "we decided that Rosboro was going to run on top."

Not everything in the old culture was conducive to the new world Rosboro found itself in. Some of its old ways had to change—in particular, the company's reticence, or, as some put it, secretiveness. Paul Cole and Ira Lowe were in the habit of informing people sparingly and reluctantly, on a "need to know" basis only.

Under the reorganized management system, that reticence had to change. No longer in their divisional silos, managers had to share information—and opinions—freely, and they do, said Dallas Davis. "That suspicion among the groups—that's pretty much gone. Now we can say, 'I think you're full of it' right to someone's face, as long as you have the facts to support your position."

Embracing openness also proved to be a good strategy in dealing with Rosboro's neighbors. In early 2004, soon after Knox Marshall was hired, a group of residents in the Fox Hollow area southwest of Eugene learned of Rosboro's plan to log 160 acres near their homes. They mounted a petition urging the company to reconsider.

Courtesy of Gordon Culbertson

Helicopter logging at Finn Rock tree farm

"We said, 'Let's go talk to them,'" said Jim Enright. Marshall talked to the Fox Hollow residents and learned that they had reservations about both the clearcutting and the aerial spraying of herbicides that would follow. After some dickering, he agreed not to aerial-spray but rather to apply the herbicides by hand, and he agreed to selectively cut some of the trees.

"This story has a happy ending," wrote Reida Kimmel, one of the Fox Hollow residents. "The company has gained the goodwill and respect of the community, and the land, which has been logged several times in the past and has a growth of young crowded trees mostly useful for pulp, is being treated in a responsible way that will improve the quality of the next timber harvest."[21]

"They didn't know what the impact was going to be, so they assumed the worst," said Marshall.

"We completely defused the situation," said Enright.

Notes

1 *Random Lengths* 2003 yearbook, p. 279.

2 *Random Lengths* 2005 yearbook, p. 302.

3 American Geophysical Union web site, http://www.agu.org/sci_soc/kobe.html, consulted Nov. 24, 2006.

4 "Carving a better city." New Orleans *Times-Picayune*, Nov. 24, 2006. http://www.nola.com/speced/ruinandrecovery/t-p/index.ssf?/speced/ruinandrecovery/articles/kobe1.html, consulted Nov. 24, 2006.

5 Jim Walsh explained it thus: "The major *keiretsu* were each centered around one bank, which lent money to the *keiretsu's* member companies and held equity positions in the companies. Each central bank had great control over the companies in the *keiretsu*, and it acted as a monitoring entity and as an emergency bail-out entity. One effect of this structure was to minimize the occurrence of hostile takeovers in Japan, because no entities could challenge the power of the banks. The banks covered up the problems of the companies in the *keiretsu*, causing the problems to grow as time went on."

6 "Asian economic turmoil: the fallout of the Yamaichi collapse." Public Broadcasting System Online Newshour Forum, Dec. 8, 1997, consulted Nov. 24, 2006. This is the consensus of most non-Asian analysts, but some leaders in Asia blamed their economic troubles on reckless currency speculation by Western investors.

7 "The Asian financial crisis of 1997-1998: Adapting U.S. intelligence and policy-making to the challenges of global economics." Working Group Report No. V, Institute for the Study of Diplomacy, Georgetown University, February 27, 2006, http://isd.georgetown.edu/Asian_Financial_Crisis_WG_Report.pdf, consulted Nov. 24, 2006.

8 Thoele, Michael. *Bohemia: The Lives and Times of an Oregon Timber Venture* (Portland: Oregon Historical Society Press, 1998), p. 580.

9 Aleshire, Ilene, "Rosboro buys mill in Vaughn," Eugene *Register-Guard*, Oct. 28, 2005.

10 *Random Lengths* 2001 newsletter, p. 258.

11 "Lumber firm cuts shifts at Springfield, Ore., stud mill." Eugene *Register-Guard*, Sept. 12, 2000.

12 Weza is willing to let Taylor keep his story, but "I don't remember that comment, and I'd think I would remember it," he said. "I never had the aspiration to be CEO of this company. VP, maybe."

13 Once Walsh was on board, Bauer decided to stay a while longer, which made his job a little ambiguous sometimes, Walsh said. "I was running sales, but Fred was still the virtual sales manager. Our lines were blurred." Bauer ended up staying for another five years.

14 Now firmly retired and living in Jacksonville, Florida, Bauer acknowledged, "I was comfortable there. I wasn't ready to go, so I stayed on."

15 Interview with Lynn Damon, April 23, 3007.

16 Dietz, Diane. "2004 'as good as it gets' for wood products industry." Eugene *Register-Guard*, Feb. 6, 2005.

17 *Random Lengths* 2005 yearbook, pp. 280-81.

18 Ibid., Thoele, p. 551.

19 Paul Cole has a high regard for Lynn Damon's fiscal acumen. "She is the wisest person—she has a feel for what the numbers mean, and she can put her finger on the really important thing."

20 Aleshire, Ilene, "Rosboro buys mill in Vaughn," Eugene *Register-Guard*, Oct. 18, 2005.

21 Eugene Natural History Society web site, http://biology.uoregon.edu/enhs/archive/apr04/apr044.html.

9

WHAT MAKES ROSBORO RUN?

"The reason we're doing a book now is because we've turned a corner. We've had the courage to change. We're not a traditional lumber company any more."

—David Weza

In Springfield's City Hall courtyard is a fountain, and on the fountain is a sculpture of a male gymnast balanced on his left hand with his feet in the air. His hand grips the top of a pole mounted on the back of a rhinoceros. Perched on the man's outstretched left foot is an owl. The rhino is made of rough-hewn concrete; man and bird are finely detailed in bronze. The statue honors Victor Tomseth, a Springfield boy who went on to become an officer in the U.S. Foreign Service, and who was one of the hostages seized in 1979 and held for 444 days in the U.S. Embassy in Tehran.[1]

The balancing gymnast might also be considered emblematic of the balancing act demanded of timber companies in towns like Springfield all up and down the West Coast. The lumber business has never been an easy way to make a living, and it hasn't gotten any easier since the Booth-Kelly Lumber Company first fired up its Springfield headrig in 1901. Decisions must be made daily, monthly, yearly, and everything comes down to a balance. Buy timber or build mills? Keep employees on when the mill is losing money, or shut down and lay them off? Invest in established machines and methods, or take a risk and create new ones? Perfect what you do well, or branch out into new products? Embrace growth, or find a size that fits you and stay there?

Rural Oregon is dotted still with expanses of broken concrete and rusted machinery and oxidized domes of abandoned wigwam burners. It is haunted still with the ghosts of small and medium-sized, locally owned and operated lumber companies that sparked into life, flamed for a while, and burned out. Names like Snellstrom, Bohemia, Hult, C.D. Johnson, Willamette Valley Lumber Co., and Robert Dollar are no longer printed on thousands of paychecks every week. A few of the men who started these businesses—Nils Hult, LaSells Stewart, George Gerlinger—are commemorated on plaques at opera houses or lecture halls or public parks, attesting to the indelible stamp of timber on the economy and culture of this wooded land.

A 1955 Lane County business directory lists eighty-five sawmills, plywood plants, and logging businesses. Today only seven on that list are still in operation. Rosboro is one of them.[2] "The others came and went," observes Fred Bauer, "but you've got to give Rosboro credit—they have the staying power."

Rosboro's Springfield operations, 1990

How has Rosboro managed to hang on, let alone operate almost continuously, through such tough times? There are both practical and philosophical answers to this question, and both have to do with an ongoing balancing act—how Rosboro has adapted to a constantly changing environment by balancing its priorities.

The most obvious practical answer to the question is expressed in one word: timber. Not every early entrepreneur was as foresighted as Whit Rosborough, who bought a whole township's worth of old-growth Douglas-fir up Quartz Creek in 1938. At 75 cents per thousand board feet, Rosborough got a good deal, but it was not the stupendous deal it might appear from today's perspective. There was plenty of timber back then.[3] The main obstacles to success for lumbermen were the difficulty of acquiring mill machinery and lack of roads and other infrastructure. All these took precious capital, and companies didn't want to tie up too much cash in a nonliquid asset that was practically as common as dirt. David Weza recalls hearing one old-timer say, "When timber gets to be a dollar a thousand, I'm out."

Over the years, following Rosborough's lead, the company has been steadfast in its policy of acquiring and maintaining its own timberlands. But it has done so judiciously, balancing its need for timber with its other needs, and being careful not to get itself overextended. That strategy may have saved the company in the shakeout of the 1980s and '90s, when many of Rosboro's contemporaries went under because they could not get enough timber to run their mills.

Moreover, Rosboro followed its investment in timberland with an early and foresighted investment in silviculture. The hiring of Dave Burwell in 1951 signaled the company's intention to be in the timber business for a long time. On the Quartz Creek tree farm, the second-generation forest is now reaching harvesting age, and the third generation is coming up fast behind it. In all, Rosboro now owns almost 80,000 acres of timberland in twelve counties. These lands furnish a substantial fraction of its raw-material needs. The rest is purchased from other private owners. The company is not counting on Forest Service timber. "We don't put any trust in the federal system," said David Weza.

Another practical explanation for Rosboro's longevity is its habit of taking well-researched, judicious risks. In 1959 the company spent $2 million—not small change in those days—building a plywood plant to take advantage of a surging market. In 1975 Rosboro installed a small-log sawmill at a time when most of the lumber from the Northwest was coming from big, valuable old-growth logs. The immediate impetus for building Mill B was that smaller logs were cheaper than big ones. As the river of old-growth logs feeding Northwest mills subsequently slowed to a trickle, the decision to go ahead with the stud mill proved extremely foresighted.

Such bets didn't always pay off immediately—the glulam plant took several years to become profitable. And sometimes, as with the Feather River sawmill, Rosboro got unlucky. But thanks to the company's balance of risk with caution, most of its big decisions have proven foresighted.

Rosboro is also getting bolder and more strategic with its financing. In 2005 Scott Nelson came on as the company's first full-fledged chief financial officer. At this writing Nelson is all of thirty-nine years old, too young to have absorbed Rosboro's Depression-era abhorrence of debt. To him, debt is morally neutral—it is a tool, neither good nor bad, that can be used in smart ways or not. Historically, financing was handled on a project-by-project basis, and it was not as efficient as it could have been. Nelson arranged for a new $225 million line of credit to enable the company to implement its growth strategy. He manages every aspect of finance and administration to be sure the company is getting the best possible deal.

Peeler blocks in front of Mill A addition

Rosboro's habit of constantly upgrading and modernizing equipment may be another factor in its long life. The company has learned the value of investing in improvements before circumstances force it to. During the first decade of the twenty-first century Rosboro replaced the machinery in Mill B piece by piece. A new planer and sorter with automatic grading, installed in January of 2007, completed the company's second major overhaul of the thirty-year-old stud mill. The company installed new production flow lines complete with two lumber sorters and all-new manufacturing equipment.

A fourth and very important factor is that Rosboro is not a huge, publicly traded corporation, but a relatively small, family-owned company. As such, while the company has always focused on profit, it does not face the same degree of shareholder pressure for short-term profits as do publicly traded companies. Most of the owner-partners, said Cole, feel strongly that the company should remain in family hands: "Once you have a bunch of [public] shareholders looking over your shoulder, you've lost control."

Family ownership is no guarantee of stability, however. The families that composed the original partnership have branched and branched again over five generations. Today Rosboro has 33 owners, holding their shares in a variety of corporations, trusts, holding companies and other arrangements intended to discourage them from selling their interest outside the family. The owners' views are represented by a board of directors composed mostly, at this writing, of second-generation owners such as Hub Lindsey and Jim Watts. As with most family-owned businesses, there is some worry that the next generation will have a weaker attachment to the business than their elders.

A log fork hoists a big load

Cole's colleague Foster "Robbie" Robinson is president of Starfire Lumber of Cottage Grove. Starfire is a good customer for Rosboro's big logs, and Robinson is a long-time colleague and admirer of Paul Cole and Rosboro Lumber Company. "There's a saying in the business," said Robinson. "The first generation builds a business, the second generation expands it, and the third generation sells it."

Family ownership has provided a relatively stable environment for the company. It has enabled it to take the long view: Rosboro can afford to manage its timberlands for sustained yield over time, letting them accumulate value on the stump. Rosboro's growing forestland portfolio lends yet more stability to the company by helping ensure a steady raw-material supply into the future. Rosboro does not generally buy timberlands for speculative purposes. Although parcels are bought and sold frequently, the objective is usually to maximize the company's supply of timber for its operations.

Feller-buncher at work at Finn Rock

A final factor is that Rosboro passed an important hurdle by successfully shifting power from one generation of managers to the next. This is a notorious stumbling block for family-owned businesses. In the opinion of the current managers, Cole has prepared the company well to survive his passing from the scene. "Paul and David did a lot of work to get us mentally ready," said Dallas Davis. "Five to seven years ago, we couldn't have set the goals we've set for ourselves" and expect to reach and exceed them. "Today we can."

Rosboro's owners, too, feel the transition has been successful and are generally happy with the current management. "When my brother promoted David Weza," said Ruth Cole Kainen, "I said, 'he's your man!' I just knew he was good. And my brother was willing to step down, though it was hard."

"It's hard for somebody like Paul to step aside," agrees Hub Lindsey. "[The transition] evolved gradually, and Paul and David cooperated well. Paul is now more a consultant than a day-to-day manager, and David has gone out and picked the best people [for his management team]. I think he's done an outstanding job, and I think the quality of that management group is outstanding."

Values

The practical facts of Rosboro's long run—plentiful timber, judicious risk-taking, family ownership, and a carefully planned transition—all are rooted in less tangible but deeply felt values that make up Rosboro's company culture. These values are difficult to articulate, and one would be hard put to prove them in a court of law, but they infuse every management decision at Rosboro, every interaction between worker and supervisor. They reveal themselves in the aphorisms that pop out of conversations with the company's principals. "Plan for the downside," Paul Cole is fond of saying. "There'll always be one." And "We're a secretive company. We don't tell our business to everybody." And "We have a saying around here: you don't have to limit yourself to 40 hours a week."

Rosboro is deeply conservative in a pragmatic way. It prizes commitment, prudence, reticence, hard data, and hard work. It abhors flamboyance, overconfidence, sentimentality, and speculation.

Even though it is one of Lane County's major employers, Rosboro has deliberately maintained a low profile to the outside world. The company "never wanted publicity," said Louise McQuin. "A company doesn't want everybody knowing its business. They are not gaudy people. They never went around tooting their horn."

"The company has always been secretive," said Nonnie Cole. "Paul doesn't let much out. He has the feeling that he doesn't want anybody to know" what he and the company are doing. Such secrecy may be a liability in the long run, she adds. "I think the whole industry could do more in telling its story. The environmental movement has made the industry into a dirty word, and the industry has not done enough to tell its side of the story."

The company views itself as a scrappy contender among the medium-to-large, regionally based forest-products companies, said Jim Enright. "Weyerhaeuser, Hampton, Stimson, Roseburg Forest Products—these are larger, and we're not in the same neighborhood. But when you get just below the large [category], we have a very good image as a financially sound, well-managed company." Rosboro must compete with those larger companies to survive, added Jim Walsh, "but we go to market with a different style. We don't go toe-to-toe with them. We find our niches."

Enright cites leadership in glulam production and nationwide markets as evidence that Rosboro is not an overgrown small company, but a lean-operating big one. "Most other companies our size, they stick to the West Coast. We like to branch out. We're not followers, we're leaders. We may not be well known among the public, but we're well known by our peers."

The company's conservative philosophy embraces an old-fashioned faith in individual talent, industry, and pluck. There was a day when the company prized like-mindedness in its employees and had a certain cold-bloodedness about getting rid of square pegs. Today's culture is more diverse, more team-oriented, and more tolerant of differing viewpoints, according to the unanimous opinion of the executive team. Nevertheless, "Rosboro is still a company that expects a lot from people," said Paul Cole. "Some people just don't fit. If they don't buy into the company's culture and philosophy, they're doomed, and if the company keeps them, it's a fool. Besides, it's demeaning to the person. Just because they don't fit here doesn't mean they won't fit elsewhere."

The company's conservative philosophy believes subordinates appreciate a demanding work environment and a clear line of authority. Sales manager Jim Enright said, "We are still somewhat old-school—you come in, you work hard, you get paid. Our employees are proud that they work for Rosboro, and not because it's an easy place to work."

The consensus among the employees is that Rosboro treats them fairly, said Ray Everts, a twenty-seven-year employee and president of Local 2750 of Lumber and Sawmill Workers Union, which represents most of Rosboro's hourly employees. Not only are the pay and benefits above average, but, Everts feels, the relationship between worker and management is one of mutual respect. "I've always liked Paul Cole, and he's always been one who liked the employees. We've got some guys here in the second and third generation of Rosboro employees."

Former head sawyer Roy Edwards, who worked for Rosboro for forty-four years, said, "I don't think I ever got up a day I didn't want to go to work." At eighty-eight, he is "a mite old to be [working] in a sawmill. But if I were capable and able, I'd love it."

Long-time employee Richard Johnson puts it simply: "All you've got to do is come to work and do your job." Johnson has worked at Rosboro forty-one years, most recently in the plywood plant. "That's what you hire on for, and that's what they want—an honest day's work."

Wes Bergen, still in his twenties, came to Rosboro two years ago from Weyerhaeuser, where, he said, "they didn't like my 'go get 'em' attitude." Willing to tackle any job, "I was showing up all the old guys." Bergen works in Mill B on the sorter. He relishes the variety and fast pace of his work: "I'm on the go all the time." He appreciates the close attention to safety, having escaped severe injury once by the blades of a fan that should have been covered. "They take [safety] fairly seriously, maybe not 100 percent but near the top. And everybody looks out for everybody else, so we don't do nothing stupid. We want to go home with all ten fingers and toes and our arms and legs attached."

The work environment is demanding, but Bergen likes it that way. He likes the competition among the shift crews to turn in the highest board footage. Recently his shift set a productivity record of 385,000 board feet for a 10-hour day. "We were flat moving," he said. "They lean on us pretty hard, but that's understandable. We have to get the job done and get the product out. I do about ten times more than what the job description calls for. The more you do, the better off you are."

The company's conservative philosophy believes a job well done ought to be its own reward, that praise ought to be administered sparingly and only when richly deserved. "My father," said Ruth Cole Kainen, "was the kind of person who never gave credit—he never complimented Paul, never told him what a good job he was doing. Others complimented Paul to me, and I told them, 'why don't you tell him yourself?'"

The company's conservative philosophy approves of well-defined roles for its employees and disapproves of fraternization between managers and workers. Praising the extensive knowledge of an assistant who helped him get his feet on the ground when he arrived in 1980, David Weza nevertheless said, "He couldn't really manage people. His approach was to work alongside them, not supervise them." Said Jim Enright: "People want to know what their boundaries are. There's a difference between respect and popularity. This is a good thing, and we don't want to lose it."

The company's conservative philosophy commits it firmly to private enterprise. Speaking of the near-endless trade disagreement between Canada and the United States over softwood lumber, Paul Cole observed, "Canada's goal is full employment, and ours is free enterprise. We have completely different values." What he did not need to say was his firm belief that the American values are superior.

The company's conservative philosophy considers that keeping people employed is its main contribution to the public good. Louise McQuin recalls the time Paul Cole served on the board of McKenzie-Willamette Hospital. "He finally had to give it up," she said, "because he was having more responsibility all the time in the company." Then someone from the hospital, probably a fundraiser, McQuin said, buttonholed him and demanded to know what he did for the community. "Well, I employ five hundred men," he told her. "I thought I was doing it."

Seizing the Future

The reader may wonder why a company that values modesty, that is secretive by its own admission, that has kept its head down and its nose to the grindstone for sixty-seven years, has chosen to reveal itself in a company biography at this moment.

For Paul Cole, the main reason is to commemorate the past. The older generation is passing from the scene, and a rich and edifying history is passing with it. He hopes Rosboro's story will instruct and inspire the younger generations of owners.

For Dallas Davis, it's to celebrate the company's survival. "Lots of companies Rosboro's size are not here today," he said. "We've weathered the bad times, and we're still here."

For David Weza, it's to help focus the attention of Rosboro's owners, managers, and employees on the vital task of seizing the future. "The reason we're doing a book now," he said, "is because we've turned a corner. We've had the courage to change. We're not a traditional lumber company any more."

Notes

1 The Political Graveyard web site, http://politicalgraveyard.com/bio/tomlison-tomsen.html.

2 Oregon Eugene Springfield Lane County Directory (1955), p. 22.

3 Paul Cole also points out the decisive role of World War II in boosting the company's fortunes a critical time, as it did those of most other Northwest lumber companies.